Series on Analysis, Applications and Computation – Vol. 9

ISAAC

Nonlinear Waves
A Geometrical Approach

Series on Analysis, Applications and Computation

Series on Analysis, Applications and Computation – Vol. 9

Nonlinear Waves
A Geometrical Approach

∘ Petar Popivanov
Institute of Mathematics and Informatics
Bulgarian Academy of Sciences, Bulgaria

∘ Angela Slavova
Institute of Mathematics and Informatics
Bulgarian Academy of Sciences, Bulgaria

World Scientific

NEW JERSEY · LONDON · SINGAPORE · BEIJING · SHANGHAI · HONG KONG · TAIPEI · CHENNAI · TOKYO

Published by

World Scientific Publishing Co. Pte. Ltd.

5 Toh Tuck Link, Singapore 596224

USA office: 27 Warren Street, Suite 401-402, Hackensack, NJ 07601

UK office: 57 Shelton Street, Covent Garden, London WC2H 9HE

Library of Congress Cataloging-in-Publication Data

Names: Popivanov, Peter R., author. | Slavova, Angela, author.

Title: Nonlinear waves : a geometrical approach / by Petar Popivanov (Bulgarian Academy of
      Sciences, Bulgaria), Angela Slavova (Bulgarian Academy of Sciences, Bulgaria).

Description: New Jersey : World Scientific, 2018. | Series: Series on analysis, applications, and
      computation ; volume 9 | Includes bibliographical references and index.

Identifiers: LCCN 2018043691 | ISBN 9789813271609 (hardcover : alk. paper)

Subjects: LCSH: Nonlinear wave equations. | Nonlinear waves. | Mathematical physics. |
      Nonlinear partial differential operators.

Classification: LCC QA927 .P6595 2018 | DDC 531/.113301515353--dc23

LC record available at https://lccn.loc.gov/2018043691

British Library Cataloguing-in-Publication Data

A catalogue record for this book is available from the British Library.

For any available supplementary material, please visit
https://www.worldscientific.com/worldscibooks/10.1142/11025#t=suppl

Printed in Singapore

Preface

This book deals with several equations of Mathematical Physics such as: nonlinear wave equation, Korteweg de Vries (KdV), Fornberg-Whitham, Vakhnenko, Benney-Luke, Kadomtsev-Petviashvili (K-P), Tzitzeica, Camassa-Holm, nonlinear Schrödinger (NLS), nonlinear Kaup-Kupershmidt, Boussinesq, Kuznetsov-Zakharov, Swift-Hohenberg ones, Manakov system, symmetrizable quasilinear hyperbolic systems arising in fluid dynamics, and many others. As is well known, different physical events are described by different mathematical models and therefore by different nonlinear Partial Differential Equations (PDE) in general. The corresponding solutions we are looking for can differ from each other in their structure as for example different polynomial perturbations of some linear PDE (say, the Schrödinger one) lead to global in time solutions, solutions exploding in finite time, solutions with different behavior at infinity etc. Thus, the mathematical machinery used in our investigations has the character of a mosaic. More precisely, we propose here the study of the above mentioned equations via purely mathematical methods such as Schwartz distribution theory and (partial) Fourier transform in different functional spaces, L_p-Lebesgue spaces and Sobolev spaces, Jacobi elliptic functions, Weierstrass functions and Legendre elliptic integrals of first, second and third kind. In order to find traveling wave solutions and to draw their profiles as well as to construct appropriate solutions with special physical meaning in laser optics, fluid dynamics or rogue waves in the oceans, at first methods originating from the ODE (ordinary differential equations) and the theory of first order nonlinear PDE (characteristics, first integrals) are adapted. Then in different forms the method of the simplest equation and its generalizations are often used. Later on in more detail two interesting and relatively powerful approaches for "solvability in solitons" of many equations of

Mathematical Physics are explained. It is all about the dressing method and the Hirota's direct method. The first one is illustrated mainly by the cubic NLS operator, while for the second one we propose several examples — KdV, Vakhnenko, Kaup-Kupershmidt equations.

This book is addressed to a broad audience including graduate students, Ph.D. students, mathematicians, physicists, engineers and specialists in the domain of PDE and Differential Geometry and their applications to the nonlinear equations of Mathematical Physics. Certainly, there are many monographs and plenty of papers on the subject based on rather complicated and technically difficult methods that make the subject hard to understand — especially for beginners and non specialists on PDE. It is worth to mention that the book is written with a stress on the mathematical point of view. There are proofs (more or less elementary) which are short and we do hope clear to the reader. Otherwise, the book would turn out into a handbook on several popular equations of Mathematical Physics equipped with formulas for their soliton, kink and exploding type solutions. This is not the aim of the authors as such formulas with small comments are widespread in the Internet and in reference and manual books.

To illustrate the functional, analytical and geometrical methods for solvability of several classes of evolution PDE and to accustom the reader to work on their own in that domain we include in the book a lot of examples on nonlinear differential equations arising in the applications in mechanics and physics. As in many cases these equations are modeling waves having different geometrical structure (traveling waves, kinks, solitons, loops, butterflies, ovals, peakons and antipeakons and others) we provide the book with about 50 figures. They represent a geometrical visualization of the waves under consideration and their interaction. Special attention is paid to the interaction of 2 (3) kinks-solitons, to the interaction kink-peakon, peakon-(anti)peakon, fluxon-(anti) fluxon and to the resonance effect when newborn waves appear.

We recall that in the classical theory of characteristics for first order nonlinear PDE the solutions can be interpreted as surfaces (the so-called integral surfaces), weaved from some initial surface transversally along the characteristic curves. We point out that in the investigation of inextensible moving curves in $\mathbf{R}^3$ the cubic NLS naturally appears. Therefore there is another link between the solutions of a special class of second order nonlinear PDE and the moving Frenet-Serret frame well known from the Euclidean theory of smooth curves in $\mathbf{R}^3$.

We omit in this book the inverse scattering method (ISM) for studying of some equations of Mathematical Physics as there are numerious monographs and popular books on the subject. Instead, we consider as simple examples several Volterra type linear integral equations in infinite intervals $(-\infty, x)$ that can be solved in explicit form by appropriate ansatz except the Laplace transform. Having those solutions and applying the well-known explicit formulas from the ISM we can express the solution of the evolution PDE corresponding to the above mentioned Volterra equation via some solution of the latter one. ISM originated from an application to the KdV equation in 1967 in a famous paper written by C. Gardner, J. Green, M. Kruskal, R. Miura and then in a 1968 paper by P. Lax.

We give below a short description of each of the chapters.

Chapter 1 deals with several simple exercises and examples of different evolution PDE that can be considered as preparatory ones for the book. Volterra type integral equations with separate variables are solved too.

A full classification and profiles of the traveling wave solutions of the Fornberg-Whitham equation are proposed in Chapter 2. Moreover, traveling wave solutions of the Vakhnenko and Benney-Luke equations are found and written in explicit form. In many cases they are expressed via Legendre's elliptic functions.

In Chapter 3 explicit formulas to the solutions of the nonlinear multidimensional Klein-Gordon, wave and Tzitzeica equations are constructed. The first order cubic hyperbolic pseudodifferential equation (Szegö equation) is solved too.

Chapter 4 deals with the interaction of peakon-antipeakon and kink-peakon solutions of the generalized Camassa-Holm equation. It is an evolution PDE containing both quadratic and cubic nonlinearities. Looking for solutions in a special form the investigation of the interaction of these waves is reduced to the study of a system of ODE having non-smooth right-hand side. We are able in some cases to obtain global first integrals of these systems and to study the interaction of the waves using methods in classical analysis.

An introduction to the dressing method in the form of Zakharov-Shabat is proposed in Chapter 5. This method is illustrated by the 1-D cubic nonlinear Schrödinger operator. The book is self-contained and all the necessary mathematical tools are formulated in detail. A geometrical interpretation of the so-called bright solitons is given in Section 5.4. An explicit formula for a special family of solutions of Kadomtsev-Petviashvili equation is found solving a Volterra type integral equation.

In Chapter 6 we describe Hirota's approach and the corresponding modifications for finding soliton type solutions of several equations of Mathematical Physics. We illustrate this direct approach by KdV equation, Boussinesq equation, Kaup-Kupershmidt equation, Kadomtsev-Petviashvili equation and by Vakhnenko equation. In the resonance case the interaction of 3 waves satisfying (K.-P.) equation gives rise to 3 newborn waves, while a rather exotic picture appears in the case of the interaction of two loop solutions of the Vakhnenko equation. Two fluxons and fluxon-antifluxon interactions (geometrical picture) are studied too.

In Chapter 7 several methods (elementary and not complicated) for finding explicit solutions of some nonlinear evolution PDE are proposed. It is about the method of the auxiliary solution (of the simplest equation) applied to the Swift-Hohenberg, Boussinesq, Boussinesq Paradigm, Manakov, cubic Schrödinger, derivative nonlinear Schrödinger, the Zakharov-Kuzhetsov and Liouville hyperbolic equations, the nonlocal PDE modeling the harmonic mode-locking laser and others. The interaction of two soliton solutions of Boussinesq equation and the rogue waves satisfying the cubic nonlinear Schrödinger equation are studied at the end of the chapter.

The first part of Chapter 8 deals with the regularity properties of the 2-D semilinear wave equation with radially smooth Cauchy data. The propagation of the wave-front sets of the solutions of fully nonlinear systems of PDE is studied in Section 8.3. To do this the machinery of the paradifferential operators is used. The corresponding regularity results are illustrated by an example from fluid mechanics.

Many results of Chapters 1, 2, 3 and some results of Chapters 7, 8 are due to the authors. The other results are due to other mathematicians and physicists and we hope they are incorporated into the book in a natural way.

Some of the results proposed here and their geometrical visualization as traveling or interacting nonlinear waves were given in a course of lectures in the Department of Mathematics at the University of Cagliari in May 2016 and May 2017.

Acknowledgments are extended to Prof. S. Piro-Vernier and Prof. A. Loi. Many thanks to T. Valchev for discussion on Chapter 5, to B. Yordanov on Chapter 1 and to Maya Markova for the preparation of the figures.

Petar Popivanov

Angela Slavova

Sofia 2018

Contents

Chapter 1

Introduction

1.1 Introduction

We propose in this Chapter several exercises and examples of different evolution PDEs that can be considered as preparatory ones for the main part of our book. The idea is moving from examples to the essence. Some of them will introduce the reader to the construction into explicit form of soliton solutions of some model systems arising in physics. The soliton is a special solitary nonlinear wave that after a collision with another solitary wave asymptotically remains unschated. More precisely, the soliton has the same profile up to a phase change. The classical method of characteristics and the propagation and interaction of nonlinear waves is considered too. In some cases the hodograph method is useful in constructing of solutions of 2×2 homogeneous quasilinear systems. This approach is also included in that chapter.

We introduce also the reader to a Direct method for finding solutions (including solitons) of the Korteweg de Vries equation by using linear integral equations. Briefly, at first one solves a system of two linear PDEs $L_1 F = 0$, $L_2 F = 0$ and then considers a Volterra's type linear integral equation $K + F + \int_x^\infty KFdz = 0$. Each solution $K(x, y, t)$ of the latter one gives a solution of the KdV equation by the formula $u = 2\frac{d}{dx}K(x, y, t)$. This way one can get at least theoretically a larger class of solutions of the KdV equation than can be found by the Inverse Scattering Transform too. The Hirota's direct method [63] and some modifications will be described and illustrated by many examples further in Chapter 6.

At the end we give a method of linearization of quasilinear PDEs by adding a new variable. The above mentioned method is illustrated by the Burgers-Hopf equation. Most parts of the examples are taken from different

1

sources: [28], [34], [38], [63], [107], [117], [138].

To begin with we shall consider the following strictly hyperbolic semilinear system of PDE:

Exercise 1.

$$\left| \begin{array}{l} \frac{\partial u}{\partial t} - \frac{\partial u}{\partial x} = 0, \; u|_{t=0} = \varphi_1 = \begin{cases} (x-1)^n, \, x \geq 1 \\ 0, \qquad\quad x \leq 1 \end{cases} \in C^{n-1} \\[2mm] \frac{\partial v}{\partial t} + \frac{\partial v}{\partial x} = 0, \; v|_{t=0} = \varphi_2(x) = \begin{cases} \varepsilon(x+1)^n, \, x \leq -1 \\ 0, \qquad\quad x \geq -1 \end{cases} \\[2mm] \frac{\partial w}{\partial t} = g(w)uv, \, w|_{t=0} = 0, u = u(t,x), v = v(t,x), w = w(t,x) \end{array} \right.$$

and $\varepsilon = 1$ for n-even, $\varepsilon = -1$ for n-odd.

The function $g \in C^\infty(w \geq 0), g(w) \geq C > 0$ and either

$$(a) \quad \int_0^\infty \frac{dw}{g(w)} < \infty \quad \text{or} \quad (b) \quad \int_0^\infty \frac{dw}{g(w)} = \infty.$$

In the case (a) $F(w) = \int_0^w \frac{d\lambda}{g(\lambda)}$ is a diffeomorphism $F : [0, \infty) \to [0, A)$, $F'(w) > 0, F(\infty) = A$, while in case (b) $F : [0, \infty) \to [0, \infty)$ is a diffeomorphism.

The only solution of $\frac{\partial u}{\partial t} - \frac{\partial u}{\partial x} = 0$ is given by

$$u = \begin{cases} (x+t-1)^n, \, x+t \geq 1 \\ 0, \qquad\qquad x+t \leq 1 \end{cases} \in C^{n-1}$$

and

$$v = \begin{cases} \varepsilon(x-t+1)^n, \, x-t \leq -1 \\ 0, \qquad\qquad\;\; x-t \geq -1 \end{cases}$$

The characteristics of the system are:

$$L_1 : x+t = C_1, L_2 ; x-t = C_2, L_3 : x = C_3.$$

Therefore there is a triple collision of L_1, L_2, L_3 at $(1,0) = P$ for $C_1 = 1$, $C_2 = 1$, $C_3 = 0$. Evidently $uv = 0$ for $t \leq 1 + |x|$, $uv = ((t-1)^2 - x^2)^n$, $t \geq 1 + |x|$, i.e. inside the angle with sides L_1, L_2 and containing the axes $\vec{1t}$. Therefore,

$$w(t,x) = F^{-1}\Big(\int_{1+|x|}^t ((\tau - 1)^2 - x^2)^n d\tau \Big)$$

for $t \geq 1 + |x|$, $w = 0$ for $0 \leq t \leq 1 + |x|$. We are looking now for the set of points $(t \geq 1 + |x|, x)$ where $G(t,x) = \int_{1+|x|}^t ((\tau - 1)^2 - x^2)^n = A$ (case (a)). Fix x; then $\frac{\partial G}{\partial t} = ((t-1)^2 - x^2)^n > 0$ for $t > 1 + |x|$, $G(1+|x|,x) = 0$, $G(t,x) \to_{t \to \infty} \infty$. Consequently, there exists a unique point $t(|x|)$, satisfying $G(t, x) = A$. Having in mind that $\frac{\partial G}{\partial x} = -2xn \int_{1+|x|}^{t}((\tau - 1)^2 - x^2)^{n-1} - sign\ x.0$ for $x \neq 0$, i.e. $G(t, x) \in C^1 (t \geq 1 + |x|)$, we conclude that $\Gamma : (t(|x|), x), t(|x|) > 1 + |x|$ is a smooth curve contained in the angle with sides L_1, L_2 and having in its interior $\vec{1t}$. Certainly, $t'(x) > 0$ for $x > 0$. Assume that there exists $\lim_{x \to \infty} \frac{t(x)}{x^\alpha} = B$, $G(t, x) = A$, $\alpha \geq 1$ (the case $\alpha < 1$ is impossible). The standard change $\tau - 1 = k|x|^\alpha$ in the integral $G(t, x)$ shows that it equals to $|x|^{2n\alpha+1} \int_{|x|^{1-\alpha}}^{\frac{t-1}{|x|^\alpha}} (k^2 - |x|^{2(1-\alpha)})^n dk$. Thus $\alpha = 1$ implies for $x \to \infty$ that $B = 1$, while if $\alpha > 1$ then $B = 0$. One easily sees that $t(0) = 1 + (A(2n+1))^{\frac{1}{2n+1}} > 1$.

If Γ possesses an oblique asymptote for $x \to \infty$ it is parallel to L_2. Geometrically, we have

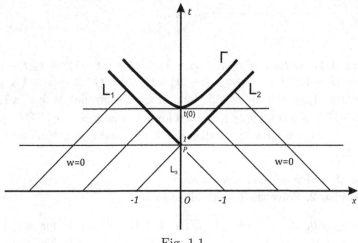

Fig. 1.1

The functions u, v have singularity at $x = 1$, $x = -1$ and they propagate along the characteristics L_1, L_2. $w = 0$ in the strip $0 \leq t < 1$, u is continuous along $\vec{1t}$. Due to the collision of 3 characteristics at $(1, 0) = P$ a new singularity was born and it propagates along $L : t = 1 + |x|$ for $x \neq 0$. It is weaker than the initial singularities as φ_1, $\varphi_2 \in C^{n-1}$, while the new born singularities belong to C^n; $w \in C^\infty$ above $L \setminus \{P\}$ and under Γ, while $w|_\Gamma = \infty$ (blow up). The explosion of w is due to the nonlinearity $g(w)$, (a).

Below we propose again a geometrical picture illustrating the behaviour of $w(t, x)$: Fig. 1.2.

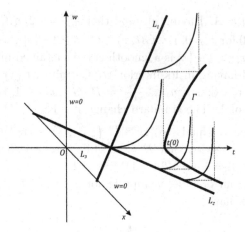

Fig. 1.2

Remark 1.1. $G(t,x) = c_1(t - |x| - 1)^{n+1}(t + |x| - 1)^n + c_2(t - |x| - 1)^{n+2}(t + |x| - 1)^{n-1} + \ldots + c_{n+1}(t - |x| - 1)^{2n+1} \in C^\infty(t \geq |x| + 1, x \neq 0)$, $\partial_t^\alpha \partial_x^\beta G(|x| + 1, x) = 0$ for $\alpha + \beta \leq n$. If φ_1, φ_2 are flat at $x = \pm 1$, say $\varphi_1 = e^{-\frac{1}{|x-1|^p}}$, $p > 0$ for $x \geq 1$, $\varphi_1 = 0$ for $x \leq 1$ and $\varphi_2 = e^{-\frac{1}{|x+1|^p}}$, $p > 0$ for $x \leq -1$, $\varphi_2 = 0$ for $x \geq -1$ then $u, v \in C^\infty$ and $w \in C^\infty$ under Γ but outside $\vec{0t}$.

Case (b) is left to the reader.

Exercise 2. Solve the hyperbolic system

$$\left|\begin{array}{ll} u_t - u_x = 0, & u|_{t=0} = 1 \text{ for } x \geq 1, \quad u|_{t=0} = 0 \text{ for } x \leq 1 \\ v_t + v_x = 0, & v|_{t=0} = 1 \text{ for } x \leq -1, v|_{t=0} = 0 \text{ for } x > -1 \\ M_1 p \equiv p_t - a p_x = q, & p|_{t=0} = 0, \qquad\qquad 0 < a < 1 \\ M_2 q \equiv q_t + a q_x = uv, & q|_{t=0} = 0 \end{array}\right.$$

Answer:

$$p = \begin{cases} \frac{(t-|x|-1)}{4a(1+a)}(|x + a(t-1)| + |x - a(t-1)|), & t \geq \frac{|x|}{a} + 1 \\ \frac{(t-|x|-1)^2}{2(1-a^2)}, & 1 + |x| \leq t \leq \frac{|x|}{a} + 1, \\ 0, & 1 + |x| \geq t \end{cases}$$

$uv = 1$, $t \geq |x| + 1$, $uv = 0, t < |x| + 1$.

Therefore, $p \in C^0(t \geq 0)$, $p \in C^1$ at the characteristic M, $M = \{t = |x| + 1, x \neq 0\}$ and $p \in C^\infty$ outside $\bar{L}$, $\bar{M}$, where the characteristic $L = \{t = \frac{|x|}{a} + 1\}$. Starting with jump type discontinuities at $(0, \pm 1)$

the characteristics $M = \{t = |x| + 1\}$ are crossing at $(1, 0)$ where new singularity appears and propagates along L, M. It is weaker (C^0) than the initial ones (jumps); $p = 0$ in $\{0 \leq t < 1\}$. Put $L_\pm = L \cap \{x > (<)0\}$.

Hint. $M_1 M_2 p = p_{tt} - a^2 p_{xx} = uv$, $p|_{t=0} = p_t|_{t=0} = 0$, $p = 0$ for $t < |x| + 1$. The solution p is given by $p(t, x) = \frac{1}{2a} \int_1^t \int_{x-a(t-\tau)}^{x+a(t-\tau)} uv d\xi d\tau$. Geometrically (see Fig. 1.3):

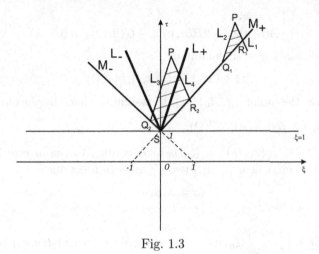

Fig. 1.3

Assume that $P(t, x)$ is arbitrary point in $\{t > 1\}$, $L_1 \| L_-$, $L_2 \| L_+$, Q_1, R_1 are the crossing points of L_2 and M_+, respectively of L_1 and M_+. Up to the factor $\frac{1}{2a}$ the value of $p(t, x)$ coincides with the area of the triangle PQ_iR_i; $Q_2 = L_3 \cap M_-$, $L_3 \| L_+$, $R_2 = L_4 \cap M_+$, $L_4 \| L_-$. $p(P)$ is $\frac{1}{2a}$ part of the area of the quadrangle PR_2Q_2S, $S = (1, 0)$ and the area of $PR_2SQ_2 =$ the area of PR_2S+ the area of SQ_2P. Knowing the coordinates of the vertices of any triangle we can find its area.

1.2 Hodograph transformation and canonical forms of linear hyperbolic PDE in $\mathbf{R}^2$

Consider the semilinear (quasilinear) PDE of hyperbolic type

$$A(\varphi_x, \varphi_t)\varphi_{xx} + 2B(\varphi_x, \varphi_t)\varphi_{xt} + C(\varphi_x, \varphi_t)\varphi_{tt} = 0, \qquad (1.1)$$

where A, B, C are at least C^2 functions of their arguments and $B^2(\xi_0, \eta_0) - AC(\xi_0, \eta_0) > 0$, $\nabla_{x,t}\varphi(x_0, t_0) = (\xi_0, \eta_0)$. We are interested in local solutions of Eq. (1.1) near (x_0, t_0) that can be written explicitly.

To do this we shall use the well-known hodograph transformation. Our first step is to reduce Eq. (1.1) to a first order system via the change near (x_0, t_0):

$$\left| \begin{array}{l} u = \varphi_x \\ v = \varphi_t, \end{array} \right. \tag{1.2}$$

i.e.

$$\left| \begin{array}{l} u_t = v_x \\ A(u,v)u_x + 2B(u,v)u_t + C(u,v)v_t = 0. \end{array} \right. \tag{1.3}$$

We shall study now the mapping

$$\Omega \ni (x,t) \to (u(x,t), v(x,t)) \tag{1.4}$$

locally near the point $(x_0, t_0) \in \Omega$, assuming that the Jacobian $j = \left| \begin{array}{cc} u_x & u_t \\ v_x & v_t \end{array} \right| (x_0, t_0) \neq 0$, $u, v \in C^2(\Omega)$.

Put $(u(x_0, t_0), v(x_0, t_0)) = (\xi_0, \eta_0)$. According to the inverse function theorem there exist near (ξ_0, η_0) two C^2 smooth functions

$$\left| \begin{array}{l} x = x(u,v) \\ t = t(u,v) \end{array} \right. \tag{1.5}$$

such that $J = \left| \begin{array}{cc} x_u & x_v \\ t_u & t_v \end{array} \right| (\xi_0, \eta_0) \neq 0$ and $(u,v) \to (x(u,v), t(u,v))$ is inverse to Eq. (1.4) near (ξ_0, η_0).

Moreover,

$$u_x = jt_v, u_t = -jx_v, v_x = -jt_u, v_t = jx_u. \tag{1.6}$$

Therefore, Eq. (1.3) takes the form

$$\left| \begin{array}{l} x_v = t_u \\ A(u,v)t_v - 2B(u,v)x_v + C(u,v)x_u = 0, \end{array} \right. \tag{1.7}$$

i.e.

$$\left| \begin{array}{l} t_u = x_v \\ t_v = \frac{2B(u,v)x_v - C(u,v)x_u}{A(u,v)}. \end{array} \right. \tag{1.8}$$

The integrability condition for $A(\xi_0, \eta_0) \neq 0$ and $C(\xi_0, \eta_0) \neq 0$, $t_{uv} = t_{vu}$ implies that the linear system Eq. (1.8) can be reduced to the linear scalar equation

$$A(u,v)x_{vv} - 2Bx_{uv} + Cx_{uu} + x_u \frac{(AC_u - A_uC)}{A} + \tag{1.9}$$

$$2x_v \frac{(A_uB - AB_u)}{A} = 0, x = x(u,v).$$

Put $[A, B] = B\frac{\partial A}{\partial u} - A\frac{\partial B}{\partial u} = -\frac{\partial}{\partial u}(\frac{B}{A})A^2$, $[C, A] = AC_u - A_u C$. Then Eq. (1.9) takes the form

$$Ax_{vv} - 2Bx_{uv} + Cx_{uu} + \frac{\partial}{\partial u}(\frac{C}{A})Ax_u - 2\frac{\partial}{\partial u}(\frac{B}{A})Ax_v = 0 \qquad (1.10)$$

i.e.

$$Ax_{vv} - 2Bx_{uv} + Cx_{uu} + \frac{1}{A}([C, A]x_u + 2x_v[A, B]) = 0. \qquad (1.11)$$

Certainly, Eq. (1.11) is linear second order strictly hyperbolic operator near (ξ_0, η_0). The equation of the characteristics of Eq. (1.11) is:

$$Cdv^2 + 2Bdudv + A(du)^2 = 0 \iff \qquad (1.12)$$

$$C(\frac{dv}{du})^2 + 2B\frac{dv}{du} + A = 0 \iff$$

$$\left(\frac{dv}{du}\right)_{1,2} = \frac{-B(u,v) \pm \sqrt{B^2(u,v) - CA(u,v)}}{C(u,v)} \Rightarrow v_{1,2} = v_{1,2}(u, C), C = const.$$

Writing Eq. (1.11) along the characteristics $\lambda = v_1$, $\mu = v_2$ we get:

$$x''_{\lambda\mu} = \Phi(\lambda, \mu, x, x'_\lambda, x'_\mu). \qquad (1.13)$$

Certainly, Eq. (1.13) is linear PDE of second order. It can be solved locally via the corresponding Riemann function [38]. In many cases we are able to solve Eq. (1.13) in an elementary way, say for $\Phi \equiv 0$ or $\Phi = k(\lambda, \mu)x_\lambda$, $\Phi \equiv const.$, etc.

We shall illustrate that approach by three examples.

Example 1. (From the theory of cylindrical shock waves [138]).

Consider 2×2 system of first order

$$\begin{aligned} \frac{\partial q}{\partial s} + q^2 \frac{\partial \Theta}{\partial \beta} = 0 \\ \frac{\partial \Theta}{\partial s} + \frac{\partial q}{\partial \beta} = 0, q = q(s, \beta), \Theta = \Theta(s, \beta). \end{aligned} \qquad (1.14)$$

According to the hodograph transformation $(s, \beta) \to (q, \Theta)$, $(q, \Theta) \to$ $(s(q, \Theta), \beta(q, \Theta))$, $q_s = j\beta_\Theta$, $j = \begin{vmatrix} q_s & q_\beta \\ \Theta_s & \Theta_\beta \end{vmatrix} \neq 0$, $q_\beta = -js_\Theta$, $\Theta_s = -j\beta_q$, $\Theta_\beta = js_q$. Therefore Eq. (1.14) can be rewritten as:

$$\begin{vmatrix} \beta_\Theta + q^2 s_q = 0 \\ \beta_q + s_\Theta = 0, \end{vmatrix} \qquad (1.15)$$

i.e. the following linear PDE arises:

$$s_{\Theta\Theta} = q^2 s_{qq} + 2q s_q. \qquad (1.16)$$

Let $s(q, \Theta) = Q(q)\tilde{\Theta}(\Theta)$. Then separating the variables in Eq. (1.16) we have

$$q^2 \frac{Q''(q)}{Q(q)} + 2q \frac{Q'(q)}{Q(q)} = \frac{\tilde{\Theta}''(\Theta)}{\tilde{\Theta}(\Theta)} = -\lambda = const.$$

i.e. $\tilde{\Theta}'' + \lambda\tilde{\Theta} = 0$, $q^2 Q'' + 2qQ' + \lambda Q = 0$ (Euler-Fuchs ODE). Taking $\lambda_n = n^2 \in \mathbf{N}_0$ we get $\tilde{\Theta}_n = k_{1n}\cos n\varphi + k_{2n}\sin n\varphi$, $Q_n = q^{\mu_n}$, where $\mu_n = -\frac{1}{2} \pm \sqrt{\frac{1}{4} - n^2}$, $Re\ \mu_n = -\frac{1}{2}$ for $n^2 \in \mathbf{N}$, $\mu_0 = \{0; 1\}$.

Example 2. (Born-Infeld equation [138]). Consider the equation $(1 - \varphi_t^2)\varphi_{xx} + 2\varphi_x\varphi_t\varphi_{xt} - (1 + \varphi_x^2)\varphi_{tt} = 0$ which is of the type Eq. (1.1) with $A = 1 - \varphi_t^2$, $B = \varphi_x\varphi_t$, $C = -(1 + \varphi_x^2)$ when $1 + \varphi_x^2 - \varphi_t^2 > 0$. Repeating the procedure described in Eq. (1.2)–Eq. (1.12) find a class of nontrivial solutions.

Hint. The linear ODE Eq. (1.13) can be reduced to $x''_{\lambda\mu} = 0$.

Example 3. Find the classical solutions of $-\varphi_x\varphi_{xt} + \varphi_t\varphi_{xx} = 0$, i.e. of Eq. (1.1) with $u = \varphi_x$, $v = \varphi_t : A = v$, $2B = -u$, $C = 0$.

Thus, $\begin{vmatrix} u_t = v_x \\ vu_x - uu_t = 0 \end{vmatrix}$, $\Rightarrow \frac{d}{dx}(\frac{u}{v}) = 0$ for $v(x_0, t_0) \neq 0$, i.e. $u(x, t) = \gamma(t)v(x, t)$ for some $\gamma \in C^2$. To fix the things let $u(x_0, t_0) \neq 0 \Rightarrow \gamma(t)(t_0) > 0$ ($\gamma(t_0) < 0$ is considered similarly). Therefore, $\gamma'v + \gamma v_t = v_x$ near (x_0, t_0). Having in mind that $\frac{dt}{\gamma(t)} = \frac{dx}{-1} = \frac{dv}{-\gamma'v}$ we obtain the independent first integrals $C_1 = x + \int \frac{dt}{\gamma(t)}$, $C_2 = \gamma v$. Answer: $v(x, t) = \frac{1}{\gamma(t)}\delta(x + \int \frac{dt}{\gamma(t)})$, $u = \gamma(t)v(x, t)$, $\delta \in C^2$, $\begin{vmatrix} \varphi_x = u \\ \varphi_t = v \end{vmatrix}$. The solution φ exists near (x_0, t_0) as $u_t = v_x = \gamma'v + \gamma v_t$. In this simple case we avoided the use of hodograph method.

Footnotes. Consider the real-valued quasilinear 2×2 homogeneous system

$$A(u, v)\begin{pmatrix} u_x \\ v_x \end{pmatrix} + B(u, v)\begin{pmatrix} u_y \\ v_y \end{pmatrix} = 0 \qquad (1.17)$$

near (x_0, y_0); put $u(x_0, y_0) = u_0$, $v(x_0, y_0) = v_0$, where $det\ A(u_0, v_0) \neq 0$, $A \in C^\infty$, $B \in C^\infty$ in a neighborhood of (u_0, v_0). Locally Eq. (1.17) is equivalent to

$$I_2\begin{pmatrix} u_x \\ v_x \end{pmatrix} + C(u, v)\begin{pmatrix} u_y \\ v_y \end{pmatrix} = 0, \qquad (1.18)$$

$$C = A^{-1}B = \begin{pmatrix} c_1(u, v) & c_2(u, v) \\ c_3(u, v) & c_4(u, v) \end{pmatrix}.$$

We assume that $c_2(u_0, v_0)c_3(u_0, v_0) \neq 0$ and $[c_1(u_0, v_0) - c_4(u_0, v_0)]^2 + 4c_2(u_0, v_0)c_3(u_0, v_0) > 0$ (implying the hyperbolicity of Eq. (1.17)).

The hodograph transformation $\Omega \ni (x, y) \to (u(x, y), v(x, y)) \in \mathbf{R}^2$ near (x_0, y_0) with $j = \begin{vmatrix} u_x & u_y \\ v_x & v_y \end{vmatrix} (x_0, y_0) \neq 0$ reduces Eq. (1.18) to

$$I_2 \begin{pmatrix} y_v \\ -y_u \end{pmatrix} + C(u, v) \begin{pmatrix} -x_v \\ x_u \end{pmatrix} = 0 \tag{1.19}$$

near (u_0, v_0).

The linear system with respect to $(x(u, v), y(u, v))$ Eq. (1.19) can be written as

$$\begin{vmatrix} y_u = c_4 x_u - c_3 x_v \equiv D(u, v, \nabla_{u,v} x) \\ y_v = -c_2 x_u + c_1 x_v \equiv E(u, v, \nabla_{u,v} x). \end{vmatrix} \tag{1.20}$$

The integrability condition $y_{uv} = \frac{\partial}{\partial v} D = y_{vu} = \frac{\partial}{\partial u} E$ implies

$$L(u) = -c_2 x_{uu} + (c_1 - c_4) x_{vu} + c_3 x_{vv} - (c_{2u} + c_{4v}) x_u + (c_{1u} + c_{3v}) x_v = 0. \tag{1.21}$$

Equation (1.21) is a linear second order hyperbolic operator with 2 families of characteristics: $c_2(dv)^2 - (c_1 - c_4)dudv + c_3(du)^2 = 0$ and $x(u, v)$ exists near (u_0, v_0). The function $y(u, v)$ can be found from Eq. (1.20) as the total differential condition $\frac{\partial}{\partial v} D = \frac{\partial}{\partial u} E$ is satisfied. Therefore, the system Eq. (1.18) is reduced to the solvability of the linear second order hyperbolic scalar equation Eq. (1.21). Conversely, solving Eq. (1.21) near (u_0, v_0) we find $x(u, v)$ and then $y(u, v)$ from Eq. (1.20). At least theoretically we can go back to the local solution $u(x, y)$, $v(x, y)$ of Eq. (1.17) near the point (x_0, y_0).

We shall continue our work proposing below several new Exercises to the reader.

1.3 Exercises on nonlinear systems of PDE

As we mentioned at the beginning of that Chapter we formulate several instructive exercises of nonlinear equations and systems. The hints are shorter and there are some difficulties to be overcome by the reader (knowledge of the elliptic functions supposed). Small problems for unaided consideration.

Exercise 3. Consider the Cauchy problem to the following first order semilinear PDE:

$$iL(u) = i\sum_{j=1}^{n} a_j \frac{\partial u}{\partial x_j} = u|u|^2, a_1 = 1, a_j \in \mathbf{R}^1 \setminus 0, \tag{1.22}$$

$$u|_{x_1=0} = \psi(x'), x' = (x_2, \ldots, x_n), u = u(x_1, x') = u(x).$$

Prove that it possesses a unique global smooth complex-valued solution for each smooth complex-valued function ψ.

Hint. After the standard linear nondegenerate change of the variables

$$\begin{vmatrix} x_1 = y_1, & y' = (y_2, \ldots, y_n) \\ x' = y_1 a' + y', a' = (a_2, \ldots, a_n) \end{vmatrix}$$

the operator L takes the form $\frac{\partial}{\partial y_1}$. Suppose that $u = A + iB$ and A, B are smooth real-valued functions. Therefore, Eq. (1.22) is reduced to the system of ODE in y_1, y' being a parameter

$$\frac{\partial A}{\partial y_1} = B(A^2 + B^2) \tag{1.23}$$

$$\frac{\partial B}{\partial y_1} = -A(A^2 + B^2)$$

and Eq. (1.23) has the first integral $A^2 + B^2 = \varphi(y') \geq 0$, φ-arbitrary, i.e. $|u|^2 = \varphi$.

Answer. The general solution of Eq. (1.22) in the y-coordinates is $u(y) = C(y')e^{-y_1|C|^2}$, where $C(y')$ is complex-valued smooth function. The solution of Eq. (1.22) is $u(x) = \psi(x' - a'x_1)e^{-ix_1|\psi(x'-a'x_1)|^2}$.

Exercise 4. Solve the Cauchy problem

$$\begin{vmatrix} iL(u) = i\sum_{j=1}^n a_j \frac{\partial u}{\partial x_j} = \bar{u}|u|^2 \\ u|_{x_1=0} = \psi(x'), a_1 = 1, a_j \in \mathbf{R}^1 \setminus 0. \end{vmatrix} \tag{1.24}$$

Answer. $A^2 - B^2 = \varphi(y')$, $y' = (y_2, \ldots, y_n)$ is a first integral of the corresponding system of ODE, similar to Eq. (1.23). Then $\frac{\partial B}{\partial y_1} = \pm\sqrt{B^2 + \varphi(y')}(\varphi(y') + 2B^2)$ and the latter ODE with separate variables with respect to y_1, y'-fixed is solved in quadratures: $y_1 + \gamma(y') = \pm \int \frac{dB}{(2B^2+\varphi)\sqrt{B^2+\varphi}}$ via the Euler substitutions.

Exercise 5. Solve the system (chain system in physics)

$$\sigma_1' = \sigma_3^2 - \sigma_2^2 + \mu_3 - \mu_2 \tag{1.25}$$
$$\sigma_2' = \sigma_1^2 - \sigma_3^2 + \mu_1 - \mu_3$$
$$\sigma_3' = \sigma_2^2 - \sigma_1^2 + \mu_2 - \mu_1,$$

where $\sigma_i = \sigma_i(x)$, μ_1, μ_2, $\mu_3 = const$.

Answer. $\gamma_i = \gamma_i(\sigma_1, \sigma_2, \sigma_3)$, $i = 1, 2$ are two functionally independent first integrals of Eq. (1.25), where $\gamma_1 \equiv \sigma_1 + \sigma_2 + \sigma_3 = C_1$, $\gamma_2 \equiv (\sigma_1 +$

$\sigma_2)(\sigma_2+\sigma_3)(\sigma_1+\sigma_3)+\mu_2(\sigma_1+\sigma_3)+\mu_1(\sigma_2+\sigma_3)+\mu_3(\sigma_1+\sigma_2)=C_2$. As it is well known, they satisfy the PDE: $(\sigma_3^2-\sigma_2^2+\mu_3-\mu_2)\frac{\partial\gamma}{\partial\sigma_1}+(\sigma_1^2-\sigma_3^2+\mu_1-\mu_3)\frac{\partial\gamma}{\partial\sigma_2}+(\sigma_2^2-\sigma_1^2+\mu_2-\mu_1)\frac{\partial\gamma}{\partial\sigma_3}=0$. Moreover, $x+const=\int R(\sigma_1,\sqrt{P_4(\sigma_1)})d\sigma_1$, $P_4(\sigma_1)$ being a 4th order polynomial of σ_1 and R-rational function. Therefore, the chain system is integrable in quadratures via Weierstrass or Legendre type elliptic functions [8], [33].

Exercise 6. Solve the following Cauchy problem with constant initial data for $t=0$ and express its solution by the elliptic functions:

$$(\partial_t+\partial_x)u_1=c_1u_2u_3 \tag{1.26}$$

$$\partial_t u_2=c_2u_1u_3$$

$$(\partial_t-\partial_x)u_3=c_3u_1u_2,$$

where c_1, c_2, c_3, $c_1c_2c_3\neq0$ are real constants.

Answer. The solutions $u_j=A_j(t)$, $1\leq j\leq3$ satisfy a system of ODE having the first integrals $\gamma_1\equiv c_2A_1^2-c_1A_2^2=D_1$ and $\gamma_2\equiv c_3A_2^2-c_2A_3^2=D_2$. Therefore, $A_2'(t)=\pm\sqrt{D_1+c_1A_2^2}\sqrt{c_3A_2^2-D_2}$. Formulas for A_2 expressed by the elliptic functions can be seen in the Handbooks [30], [33], [120].

Exercise 7. [63] (interaction of 3 waves). Consider the system:

$$L_1(u_1)=\frac{\partial u_1}{\partial t}+c_1\frac{\partial u_1}{\partial x}=\varepsilon_1u_2u_3,\varepsilon_1^2=1$$
$$L_2(u_2)=\frac{\partial u_2}{\partial t}+c_2\frac{\partial u_2}{\partial x}=\varepsilon_2u_1u_3,\varepsilon_2^2=1 \tag{1.27}$$
$$L_3(u_3)=\frac{\partial u_3}{\partial t}+c_3\frac{\partial u_3}{\partial x}=\varepsilon_3u_1u_2,\varepsilon_3^2=1,$$
$$c_1,c_2,c_3\in\mathbf{R}^1\setminus0,c_1\neq c_2\neq c_3\neq c_1.$$

a). Construct a traveling wave solution of Eq. (1.27)

b). Find a more general solution of Eq. (1.27) than the traveling waves.

Answer. a). Let $u_i=u_i(x-vt)$ $v\neq c_i$, $i=1,2,3$. Then $u_i(\xi)$, $i=1,2,3$, $\xi=x-vt$ satisfy a system of ODE, solved in the previous Exercise 6. For appropriate constants a, b, c : $u_3=c\,sech\,\xi$, $u_2=b\,sech\,\xi$, $u_1=a\,tgh\,\xi$.

Hint to b). ([63], [143]). Put $\eta_2=p_2(x-c_2t)$, $\eta_3=p_3(x-c_3t)$, $L_2(\eta_2)=0$, $L_3(\eta_3)=0$, where the parameters p_2, p_3 satisfy the relation: $(c_1-c_2)p_2=(c_1-c_3)p_3\neq0$ and look for the solutions u_2, u_3 of the form:

$$u_2=\frac{a_{21}e^{\eta_2}}{1+a_{02}e^{2\eta_2}+b_{02}e^{2\eta_3}},u_3=\frac{a_{31}e^{\eta_3}}{1+a_{02}e^{2\eta_2}+b_{02}e^{2\eta_3}};$$

a_{21}, a_{31} are arbitrary constants, while a_{02}, b_{02} will be found later on. As $u_2u_3=\frac{a_{21}a_{31}e^{\eta_2+\eta_3}}{(1+a_{02}e^{2\eta_2}+b_{02}e^{2\eta_3})^2}$ one can expect that u_1 exists in the form $u_1=$

$\frac{a_{22}e^{\eta_2+\eta_3}}{1+a_{02}e^{\eta_1}+b_{02}e^{\eta_2}}$. Substitute the expressions for u_1, u_2, u_3 in the system Eq. (1.27) and solve the 3×3 nonlinear algebraic system with the unknown quantities a_{22}, a_{02}, b_{02} that arise.

Answer. $a_{22} = \varepsilon_1 \frac{a_{21}a_{31}}{2(c_1-c_2)p_2}$, $a_{02} = \varepsilon_1\varepsilon_3\frac{a_{21}^2}{4(c_1-c_2)(c_2-c_3)p_2^2}$, $b_{02} = -\varepsilon_1\varepsilon_2\frac{a_{31}^2}{4(c_1-c_3)(c_2-c_3)p_3^2}$.

Exercise 8. Solve the Cauchy problem in the plane

$$\frac{\partial u}{\partial t} + c_1\frac{\partial u}{\partial x} = c_3uv, u = u(x,t), v = v(x,t)$$
$$\frac{\partial v}{\partial t} + c_2\frac{\partial v}{\partial x} = c_4uv, c_1,c_2,c_3,c_4 \in \mathbf{R}^1 \setminus 0 \qquad (1.28)$$
$$u(0,x) = u_0(x), v(0,x) = v_0(x), u_0,v_0 \in C^0(\mathbf{R}^1).$$

Hint. Consider the following two cases: a). $c_1 \neq c_2$, b). $c_1 = c_2$.

a). Make the change $z = x - c_1t$, $w = x - c_2t$, $c_1 \neq c_2$ in Eq. (1.28) and reduce it to

$$(c_1 - c_2)\frac{\partial u}{\partial w} = c_3uv$$
$$(c_2 - c_1)\frac{\partial v}{\partial z} = c_4uv \qquad (1.29)$$
$$u|_{z=w} = u_0(z), v|_{z=w} = v_0(z).$$

Find a solution of Eq. (1.29) having the form:

$$u = \frac{A_1(z)}{B_1(z) + B_2(w)}, v = \frac{A_2(w)}{B_1(z) + B_2(w)}, \qquad (1.30)$$

where the unknown functions A_1, A_2, B_1, B_2 are C^1 smooth and $B_1(z) + B_2(w) \neq 0$. The substitution of Eq. (1.30) in Eq. (1.29) will give that $u = \frac{(c_1-c_2)B_1'(z)}{c_4(B_1(z)+B_2(w))}$, $v = \frac{(c_2-c_1)B_2'(w)}{c_3(B_1(z)+B_2(w))}$. As $u|_{z=w} = u_0(z)$ and $v|_{z=w} = v_0(z)$ the function $B_1(z) + B_2(z) = De^{\int_0^z \frac{c_4u_0(\lambda)-c_3v_0(\lambda)}{c_1-c_2}d\lambda}$, $D \neq 0$, $D = $ const., i.e. $B_1(z) = \frac{c_4D}{c_1-c_2}\int_0^z u_0(\lambda)e^{\int_0^\lambda \frac{c_4u_0(\mu)-c_3v_0(\mu)}{c_1-c_2}d\mu}d\lambda+c_5$, while $B_2(z) = De^{\int_0^z \frac{c_4u_0(\lambda)-c_3v_0(\lambda)}{c_1-c_2}d\lambda} - B_1(z)$ for each z etc. The solution of Eq. (1.28) are expressed by quadratures.

b). $c_1 = c_2 \neq 0$. Make the change $\begin{vmatrix} z = x - c_1t \\ x = x \end{vmatrix}$ in Eq. (1.28) and reduce it to the system

$$c_1\frac{\partial u}{\partial x} = c_3uv$$
$$c_1\frac{\partial v}{\partial x} = c_4uv. \qquad (1.31)$$

Show that $\begin{vmatrix} u = \frac{c_3}{c_4}v + a(z) & \text{for arbitrary } a(z) \in C^1 \\ \frac{\partial v}{\partial x} = \frac{1}{c_1}v(c_3v + c_4a(z)) \end{vmatrix}$, z being a para-meter.

The above written Bernoulli ODE has the following general solution: $\frac{1}{v(x,z)} = K(z)e^{-\frac{c_4}{c_1}a(z)x} - \frac{c_3}{c_4 a(z)}$, where $a(z) \neq 0$ and the smooth function $K(z)$ is arbitrary up to the following restriction: $\frac{c_4}{c_3}a(z)K(z)e^{-\frac{c_4}{c_1}a(z)x} \neq 1$.

Exercise 9. [28] Solve the system

$$u' + 2uv = 0, u = u(t), v = v(t)$$
$$v' - u^2 + v^2 + 1 = 0. \tag{1.32}$$

Answer: $\gamma = (u-d)^2 + v^2 - d^2 + 1 = 0$ is a first integral of Eq. (1.32); $d = const$. For $d > 1$ the phase curves are circles. The stationary points of Eq. (1.32) are $u = \pm 1$, $v = 0$. The system describes the oscillation of the harmonic oscillator with given frequency and period T. In polar coordinates $\begin{vmatrix} u = d + r\cos\alpha \\ v = r\sin\alpha \end{vmatrix}$, $d^2 = r^2 + 1$. From the first equation of Eq. (1.32) we then obtain that $\alpha' = 2(d + r\cos\alpha) = 2u$ and therefore $\int_t^{t+2\pi} u dt = \pi$ On the other hand, $\int \frac{d\alpha}{d+r\cos\alpha} = 2\int \frac{dy}{d+r+(d-r)y^2} = 2arctg(\sqrt{\frac{d-r}{d+r}}y)$ after the change $y = tg\frac{\alpha}{2}$ in the first integral. Assuming that $\alpha(0) = 0$ we get $arctg(\sqrt{\frac{d-r}{d+r}}tg\frac{\alpha}{2}) = t \Rightarrow \alpha = 2arctg((d+r)tg\ t)$ as $(d-r)(d+r) = 1$.

1.4 Linear Volterra equations and evolution PDEs

We shall discuss briefly the link between a special class of Volterra type integral equations — the so-called Gelfand-Levitan-Marchenko equations and the famous KdV equation

$$u_t + 6uu_x + u_{xxx} = 0. \tag{1.33}$$

To construct soliton type solution of Eq. (1.33) a direct method will be used. "Direct method" means that the derivative of each solution of the above mentioned integral equation with a kernel satisfying two appropriate linear PDEs turns out to verify Eq. (1.33). This way the application of the Inverse Scattering Transform is avoided and the construction is simplified. Certainly, there are difficulties in finding the above mentioned two linear PDEs. The roots of that approach can be found in [4], [147]. We shall rely heavily on the book [5] where the following result was proved in details.

Theorem 1.1. ([5])

Consider the linear integral equation

$$K(x, y, t) = F(x, y, t) + \int_x^\infty K(x, z, t) F(z, y, t) dz \qquad (1.34)$$

and assume that its kernel $F(x, y, t)$ satisfies the linear PDEs:

$$L_1 F = (\partial_x^2 - \partial_y^2) F(x, y, t) = 0 \qquad (1.35)$$
$$L_2 F = (\partial_t + (\partial_x + \partial_y)^3) F(x, y, t) = 0.$$

Then $u(x, t) = 2\frac{d}{dx}(K(x, x, t))$ is a solution of Eq. (1.33).

Theorem 1.1 enables us to find into explicit form soliton type solutions of Eq. (1.33). In fact $L_1 F = 0 \Rightarrow F = F(x + y, t)$ or $F = F(x - y, t)$. We shall deal here with $F(x, y, t) = F(x + y, t)$. Moreover, we shall split the variables: $F(x+y, t) = c(t)d(x+y)$. From the second equation of Eq. (1.35) we get

$$\frac{c'(t)}{8c(t)} = \frac{-d'''(z)}{d(z)} = k^3 = const, z = x + y, \qquad (1.36)$$

i.e. we can take $d(z) = e^{-kz}$, $c(t) = -c_0^2 e^{8k^3 t} = -(c_0 e^{4k^3 t})^2 = -\tilde{c}^2(t)$. Due to the linearity of Eq. (1.35) the function

$$F = -\sum_{j=1}^N \tilde{c}_j^2(t) e^{-k_j(x+y)}, \tilde{c}_j = c_{0j} e^{4k_j^3 t} \qquad (1.37)$$

satisfies Eq. (1.35), $k_1 < k_2 \ldots < k_N$ and $k_i + k_j \neq 0$, $i, j = 1, \ldots, N$.

Then the Eq. (1.34) takes the form

$$K(x, y, t) = -\sum_{i=1}^N \tilde{c}_i^2(t) e^{-k_i(x+y)} - \sum_{i=1}^N \int_x^\infty \tilde{c}_i^2 e^{-k_i(z+y)} K(x, z, t) dz. \quad (1.38)$$

Certainly, Eq. (1.38) is an integral equation with degenerate kernel. It is convenient then to look for thee unknown function $K(x, y, t)$ as:

$$K(x, y, t) = \sum_{i=1}^N \tilde{c}_i \psi_i(x) e^{-k_i y}, \qquad (1.39)$$

$\psi_i(x)$ being unknown functions too.

Putting Eq. (1.39) in Eq. (1.38) and equalizing the coefficients in front of $e^{-k_i y}$ we get the linear algebraic system:

$$\psi_i(x) = -\tilde{c}_i e^{-k_i x} - \sum_{j=1}^n \tilde{c}_i \tilde{c}_j \frac{\psi_j(x)}{k_i + k_j} e^{-(k_i + k_j) x}, 1 \leq i \leq N. \qquad (1.40)$$

Introduce the matrix $C = \left|\left| \frac{\tilde{c}_i \tilde{c}_j}{k_i + k_j} e^{-(k_i + k_j)x} \right|\right|_{i,j=1}^{N}$ and rewrite Eq. (1.40) into matrix form

$$(I + C)\psi = D, \tag{1.41}$$

where $\psi = \begin{pmatrix} \psi_1 \\ \cdot \\ \cdot \\ \cdot \\ \psi_N \end{pmatrix}$, $D = - \begin{pmatrix} \tilde{c}_1 e^{-k_1 x} \\ \cdot \\ \cdot \\ \cdot \\ \tilde{c}_N e^{-k_N x} \end{pmatrix}$, $c_{ij} = \frac{\tilde{c}_i \tilde{c}_j}{k_i + k_j} e^{-(k_i + k_j)x}$, C is real-valued symmetric matrix: $C = C^t$, where C^t is the transposed matrix of C. One can see that C is positively definite, i.e. $\sum_{i,j=1}^{N} c_{ij} \lambda_i \lambda_j > 0$ for each $\lambda = (\lambda_1 \ldots \lambda_N) \neq 0$.

Thus, $det(I + C) > 0$; $\Delta = det(I + C)$.

According to Cramer's rule the solutions of Eq. (1.41) are:

$$\psi_i(x) = -\frac{\sum_{j=1}^{N} \tilde{c}_j e^{-k_j x} Q_{ji}}{det(I + C)}, \tag{1.42}$$

Q_{ji} being the cofactors of the elements of the matrix $(I + C)$. As it is well known from the linear algebra

$$\Delta = \sum_{j=1}^{N} \left(\delta_{ij} + \tilde{c}_i \tilde{c}_j \frac{e^{-x(k_i + k_j)}}{k_i + k_j} \right) Q_{ji} \tag{1.43}$$

and δ_{ij} is the Kronecker symbol. Thus, Eq. (1.39) implies

$$K(x, x, t) = -\Delta^{-1} \sum_{i=1}^{N} \sum_{j=1}^{N} \tilde{c}_i \tilde{c}_j e^{-(k_i + k_j)x} Q_{ji}. \tag{1.44}$$

Having in mind that $\frac{d}{dt}\Delta(x)$ is a sum of the determinants of the matrices in which only one column is differentiated we get:

$$\frac{d}{dt}\Delta = -\sum_{i=1}^{N} \sum_{j=1}^{N} \tilde{c}_i \tilde{c}_j e^{-(k_i + k_j)x} Q_{ji}.$$

Therefore,

$$K(x, x, t) = \Delta^{-1} \frac{d}{dx}\Delta = \frac{d}{dx} ln\Delta$$

and the solution u of Eq. (1.33) is given by the formula

$$u(x, t) = 2\frac{d}{dx}(K(x, x, t)) = 2\frac{d^2}{dx^2} ln\Delta, \tag{1.45}$$

where Δ is written in Eq. (1.43) and Q_{ji} are the cofactors of $(I + C)$.

Example. Let $N = 2$. Then

$$\Delta = \begin{vmatrix} 1 + \frac{\tilde{c}_1^2}{2k_1}e^{-2k_1x} & \frac{\tilde{c}_1\tilde{c}_2e^{-(k_1+k_2)x}}{k_1+k_2} \\ \frac{\tilde{c}_1\tilde{c}_2e^{-(k_1+k_2)x}}{k_1+k_2} & 1 + \frac{\tilde{c}_2^2}{2k_2}e^{-2k_2x} \end{vmatrix}$$

$\tilde{c}_1 = c_{01}e^{4k_1^3t}$, $\tilde{c}_2 = c_{02}e^{4k_2^3t}$ and consequently $\Delta = 1 + e^{\eta_1} + e^{\eta_2} + e^{\eta_1+\eta_2+A_{12}}$,
$\eta_1 = -2k_1(x - 4k_1^2) + \ln c_{01}^2$, $\eta_2 = -2k_2(x - 4k_2^2) + \ln c_{02}^2$, $e^{A_{12}} = (\frac{k_1-k_2}{k_1+k_2})^2$.

1.5   Concluding remarks

We give here a modification of the construction developed in the paper
[39]. It enables us to give a link between the solutions of some first order
quasilinear PDE and some higher order linear PDE in three variables. Thus,

$$L_1(u) = u_t + P(u)u_x = 0, \tag{1.46}$$

where $(t, x) \in \mathbf{R}^2$ and $P(u)$ is a polynomial of degree m. To simplify
the things we assume that u is analytic function. As we know, Eq. (1.46)
equipped with analytic initial data $u|_{t=0} = u_0(x)$ possesses (at least locally)
a unique analytic solution $u(t, x)$. It can be found via the first integrals of
Eq. (1.46): $u = C_1$, $-tP(u) + x = C_2$, $u = u_0(x - tP(u))$. Equation (1.46)
can have blow up, i.e. in some point (t_0, x_0) the solution u is continuous
but $|u_x(t_0, x_0)| + |u_t(t_0, x_0)| = \infty$. Define now the function

$$V(t, x, a) = \frac{e^{av(t,x)} - 1}{a}, a \in \mathbf{R}^1 \setminus 0, v \in C^1. \tag{1.47}$$

Evidently, $\lim_{a\to 0} V = v(t, x)$, $\frac{\partial V}{\partial x} = v_x e^{av}$, $\frac{\partial V}{\partial t} = v_t e^{av}$, $\frac{\partial V}{\partial a} = \frac{1+e^{av}(av-1)}{a^2} \to_{a\to 0} \frac{v^2}{2}$,

$$L_2(V) = \frac{\partial V}{\partial t} + P\left(\frac{\partial}{\partial a}\right)\frac{\partial V}{\partial x} = e^{av}(v_t + P(v)v_x). \tag{1.48}$$

Thus, $L_2(V) = 0 \iff L_1(v) = 0$.

Therefore, each classical solution of Eq. (1.46) gives raise to classical
solution of the linear PDE:

$$\frac{\partial U}{\partial t} + P\left(\frac{\partial}{\partial a}\right)\frac{\partial U}{\partial x} = 0. \tag{1.49}$$

$|\nabla U(t_0, x_0, a)| = \infty$ for every a, while $\frac{\partial U}{\partial a}$ is continuous at (t_0, x_0), $\forall a$.

According to Theorem 5.1.3 from Hörmander [66] Eq. (1.49) equipped with boundary data $U - \varphi = 0$ for $x = 0$, $\partial_a^k (U - \varphi) = 0$, $k < m$, for $a = 0$, $\varphi(t, x, a)$ analytic, possesses locally a unique analytic solution $U(t, x, a)$.

Proposition 1.1.

Consider the Cauchy problem

$$\frac{\partial U}{\partial t} + P\left(\frac{\partial}{\partial a}\right)\frac{\partial U}{\partial x} = 0,$$

$U - \varphi = $ *for $x = 0$, $\partial_a^k (U - \varphi) = 0$, $k < m$ for $a = 0$, φ being analytic in (t, x, a). Its unique analytic solution $U(t, x, a)$ generates a solution of the quasilinear PDE Eq. (1.46) if $\frac{\partial}{\partial a}\frac{ln(1+aU)}{a} \equiv 0$, $\forall a$ i.e. if*

$$a^2 \frac{\partial U}{\partial a} = ln(1 + aU)(1 + aU), \quad for \ each \ a \neq 0, t, x. \tag{1.50}$$

Proof. Put $V(t, x, a) = \frac{ln(1+aU)}{a} \Rightarrow lim_{a \to 0} V(t, x, a) = U(t, x, 0)$. Then $\frac{\partial V}{\partial a} \equiv 0 \iff V(t, x, a) = V(t, x, 0) = U(t, x, 0)$. Define $u(t, x) = U(t, x, 0) \Rightarrow u(t, x) = \frac{ln(1+aU)}{a} \iff U = \frac{e^{au(t,x)} - 1}{a}$ and the identity Eq. (1.48) completes the proof. $\square$

Conclusion: The classical solutions of the linear PDE Eq. (1.49) under the additional condition Eq. (1.50) generate by the formula $U(t, x, 0) = u(t, x)$ solutions of Eq. (1.46). Equation (1.50) is Euler type nonlinear first order ODE with respect to the variable a:

$$a\frac{\partial U}{\partial a} - U = \sum_{k=3}^{\infty} \frac{(-1)^k}{k(k-1)} a^{k-1} U^k + \frac{aU^2}{2}.$$

We shall illustrate Proposition 1.1 by the following four examples of initial value problems to the Burgers-Hopf equations, where $(t, x) \in \mathbf{R} \times \mathbf{R}$:

$$\begin{aligned}
\partial_t u + u\partial_x u &= 0, & u(0, x) &= u_0(x), \text{ (A)} \\
\partial_t u + u^2\partial_x u &= 0, & u(0, x) &= u_0(x), \text{ (B)} \\
\partial_t u + u\partial_x u + u &= 0, & u(0, x) &= u_0(x), \text{ (C)} \\
\partial_t u + P(u)\partial_x u &= 0, & u(0, x) &= u_0(x). \text{ (D)}
\end{aligned}$$

To linearize the above equations, we introduce a new variable $y \in \mathbf{R}$ and denote $D_y = -i\partial_y$. The new unknown and its Fourier transform with respect to x are defined, respectively, by

$$U(t, x, y) = \frac{e^{iyu(t,x)} - 1}{iy} \quad \text{and} \quad \tilde{U}(t, \xi, y) = \int U(t, x, y)e^{-i\xi x} dx.$$

It is easy to see that $U(t, x, y)$ are solutions to the following equations:

$$\partial_t U + D_y \partial_x U = 0, \qquad U(0, x, y) = \frac{e^{iyu_0(x)} - 1}{iy}, \text{ (AL)}$$

$$\partial_t U + D_y^2 \partial_x U = 0, \qquad U(0, x, y) = \frac{e^{iyu_0(x)} - 1}{iy}, \text{ (BL)}$$

$$\partial_t U + D_y \partial_x U + i D_y y U = 0, \; U(0, x, y) = \frac{e^{iyu_0(x)} - 1}{iy}, \text{ (CL)}$$

$$\partial_t U + P(D_y) \partial_x U = 0, \qquad U(0, x, y) = \frac{e^{iyu_0(x)} - 1}{iy}. \text{ (DL)}$$

Such linear equations with constant coefficients are readily solved by means of the Fourier transform. In fact, we find that $\tilde{U}(t, \xi, y)$ satisfy

$$\partial_t \tilde{U} + \xi \partial_y \tilde{U} = 0, \qquad \tilde{U}(0, \xi, y) = \int \frac{e^{iyu_0(x)} - 1}{iy} e^{-i\xi x} dx,$$

$$\partial_t \tilde{U} - i\xi \partial_y^2 \tilde{U} = 0, \qquad \tilde{U}(0, \xi, y) = \int \frac{e^{iyu_0(x)} - 1}{iy} e^{-i\xi x} dx,$$

$$\partial_t \tilde{U} + (\xi + y)\partial_y \tilde{U} + \tilde{U} = 0, \; \tilde{U}(0, \xi, y) = \int \frac{e^{iyu_0(x)} - 1}{iy} e^{-i\xi x} dx,$$

$$\partial_t \tilde{U} + i\xi P(D_y)\tilde{U} = 0, \qquad \tilde{U}(0, \xi, y) = \int \frac{e^{iyu_0(x)} - 1}{iy} e^{-i\xi x} dx.$$

Now we have only two variables, t and y, as ξ is just a (large) parameter. It is possible to use the method of characteristics in (AL), (CL) and the Fourier method in (BL), (DL).

Our goal is to obtain asymptotics for $\tilde{U}(t, \xi, 0) = \tilde{u}(t, \xi)$ as $|\xi| \to \infty$. Below we study the case (A) only.

The method of characteristics yields

$$\tilde{U}(t, \xi, y) = \tilde{U}(0, \xi, y - t\xi), \tilde{U}(0, \xi, z) = \int \frac{e^{izu_0(x)} - 1}{iz} e^{-i\xi x} dx.$$

Hence

$$\tilde{U}(t, \xi, y) = \int \frac{e^{i(y - t\xi)u_0(x)} - 1}{i(y - t\xi)} e^{-i\xi x} dx$$

and

$$\tilde{u}(t, \xi) = \frac{i}{t\xi} \int (e^{-it\xi u_0(x)} - 1)e^{-i\xi x} dx.$$

It is not very easy to estimate the latter integral as $|\xi| \to \infty$. We assume that

$$|\partial_x^k u_0(x)| \le C_k(1 + |x|)^{-d_k}, k = 0, 1, 2.$$

Let us split the integral for $\tilde{u}(t, \xi)$ into two parts using a parameter $T = T(t, \xi)$ and cutoff function

$$\kappa \in C_0^\infty, \kappa(x) = 1 \text{ if } |x| \le 1.$$

The results are

$$\tilde{u}(t, \xi) = \frac{i}{t\xi} \int (e^{-it\xi u_0(x)} - 1)\kappa(x/T)e^{-i\xi x}dx$$

$$+\frac{i}{t\xi} \int (e^{-it\xi u_0(x)} - 1)(1 - \kappa(x/T))e^{-i\xi x}dx = I_1 + I_2.$$

From $|e^{it\xi u_0(x)} - 1| \le |t\xi u_0(x)|$, we estimate trivially the second part:

$$|I_2| \le CT^{-d_0+1}.$$

In the first part I_1 we also have a trivial estimate

$$|\frac{i}{t\xi} \int \kappa(x/T)e^{-i\xi x}dx| \le C_N|t\xi|^{-1}T^{1-N}|\xi|^{-N}.$$

Thus, it remains to study

$$\frac{i}{t\xi} \int e^{-i\xi(tu_0(x)+x)}\kappa(x/T)dx.$$

If $|t\partial_x u_0(x) + 1| \ge \varepsilon$, then this integral is bounded by $C_{\varepsilon,N}T|\xi|^{-N}$ for all N. If there are finitely many $x_k = x_k(t)$, such that

$$t\partial_x u_0(x_k) + 1 = 0, \partial_x^2 u_0(x_k) \ne 0,$$

then according to the stationary phase method [67] the integral is

$$\sum_k \frac{i}{t\xi} \frac{(2\pi)^{1/2}}{|\xi\partial_x^2 u_0(x_k)|^{1/2}} e^{-i\xi(tu_0(x_k)+x_k)-i(\pi/4)sgn(\xi\partial_x^2 u_0(x_k))} + O(T|\xi|^{-5/2}).$$

Of course the optimal choice of T will be

$$T = |\xi|^\delta, \delta = \delta(d_0).$$

In the remaining cases I_1 is too complicated to study. That is why we stop here.

Chapter 2

Traveling waves and their profiles

2.1 Introduction

1. In studying the wave breaking the following nonlinear dispersive equation was proposed by Fornberg and Whitham:

$$u_t - u_{xxt} + u_x + uu_x = uu_{xxx} + 3u_x u_{xx}. \tag{2.1}$$

Moreover, in [47] (see also [138]) a soliton-peakon solution of Eq. (2.1) having the form $u = Ae^{-\frac{1}{2}|x - \frac{4}{3}t|}$, $A = const.$ was found. Recently there is a renewal of the interest with respect to this equation (see [70], [148]). More precisely, kinks-like, solitons-peakons and periodic cusp wave solutions have been constructed. The authors of [148] are looking for traveling wave solutions of Eq. (2.1) by reducing the corresponding second order ordinary differential equation (ODE) to an autonomous first order system in the plane. Having found a first integral of that system they obtain its topological phase portrait. The observations that usually soliton type solutions correspond to homoclinic orbits of the system under consideration, kinks correspond to heteroclinic orbits and periodic traveling wave solutions correspond to periodic orbits enable the authors of [148] to propose different expressions for the solutions. In some cases explicit formulas are also given.

2. We propose below another approach for investigating the same problem. We reduce our 3rd order nonlinear ODE to a (degenerate) first order ODE with separate variables and we apply to it our Theorems 3.1, 3.2 from [110]. This way, a full description of the set of solutions, their singularities and their behavior near to the singular points (peaks, cusps) is given. We find here into explicit form (via elementary functions) or into implicit form (via elliptic functions) different solutions of Eq. (2.1). As it is evident, in those cases each solution can be written in the form of definite integral.

To simplify the things we shall work with indefinite integrals too. Certainly, the solution will depend then on some constant A. In general, the solutions are not unique. For example, $4\psi^2(\psi')^2 = 1$, $\psi(0) = 0$ possesses the solutions $\psi_{1,2} = \pm|\xi|^{1/2}$, which are non-smooth at 0 but satisfy the equation everywhere. (See also [49].) The non-unicity will play important role in finding of solutions of special type by combining two or more other solutions.

3. The Vakhnenko nonlinear evolution equation

$$\frac{\partial}{\partial x}\left(\frac{\partial}{\partial t} + u\frac{\partial}{\partial x}\right)u + u = 0 \qquad (2.2)$$

was discussed for the first time in [130]. It describes short-wave perturbations in a relaxing medium when the equations of motion are closed by the dynamic equation of state. The function u is the dimensionless pressure. In a relaxing medium, neglecting nonlinear effects, weak short waves obey the linear Klein-Gordon equation. On the other hand, one must take into account the nonlinearity caused by the wave propagation rate dependence on the amplitude. This way, after certain transformations — factorization and shift in space with small perturbations velocity, one can obtain the Eq. (2.2). Our aim here is to find out traveling wave solutions of Eq. (2.2) proposing explicit forms to the corresponding solutions and giving geometrical visualization (interpretation) of these waves.

When looking for traveling wave solutions both Fornberg-Whitham and Vakhnenko equations are reduced to some first order linear ODE. This technical link between the equations under consideration motivates their insertion in the same chapter.

2.2 Preliminary notes on the traveling wave solutions of the Fornberg-Whitham equation

1. To begin with let $u = u(t, x) = \varphi(x - ct)$, $c = const$ satisfy Eq. (2.1). Put $\xi = x - ct$ and substitute u in Eq. (2.1). Obviously,

$$-c\varphi' + c\varphi''' + \varphi' + \varphi\varphi' = \varphi\varphi''' + 3\varphi'\varphi''. \qquad (2.3)$$

Having in mind that $(\varphi\varphi'')' = \varphi\varphi''' + \varphi'\varphi''$ and integrating Eq. (2.3) with respect to ξ we get:

$$c\varphi'' + \varphi(1 - c) + \frac{1}{2}\varphi^2 = \varphi\varphi'' + (\varphi')^2 - g, g = const \qquad (2.4)$$

i.e.

$$\varphi''(\varphi - c) = \frac{1}{2}\varphi^2 + \varphi(1 - c) - (\varphi')^2 + g. \qquad (2.5)$$

The standard change of the unknown function φ in Eq. (2.5): $\varphi' = p(\varphi) \Rightarrow \varphi'' = p\frac{dp}{d\varphi} = \frac{1}{2}\frac{d}{d\varphi}p^2$ and the substitution $q(\varphi) = p^2(\varphi)$ lead to:

$$(\varphi - c)\frac{dq}{d\varphi} = \varphi^2 + 2\varphi(1 - c) - 2q + 2g. \qquad (2.6)$$

Certainly, Eq. (2.6) is a linear ODE with respect to the function $q(\varphi)$. Thus,

$$q(\varphi) = \frac{1}{(\varphi - c)^2}\left[\bar{C}_1 + \int(\varphi^2 + 2\varphi(1 - c) + 2g)(\varphi - c)d\varphi\right], \qquad (2.7)$$

where $\bar{C}_1 = const.$ Put $r_2(\varphi) = \varphi^2 + 2\varphi(1 - c) + 2g$.
Then

$$q(\varphi) = \frac{1}{(\varphi - c)^2}\left[\bar{C}_1 + \int r_2(\varphi)(\varphi - c)d\varphi\right] \qquad (2.8)$$

and $q(\varphi) \geq 0$ on some subinterval of the real line $\mathbf{R}_\varphi^1$. After two integrations by parts we rewrite Eq. (2.8) in the form:

$$q(\varphi) = \frac{\bar{C}_1}{(\varphi - c)^2} + \tilde{r}_2(\varphi), \qquad (2.9)$$

where the quadratic polynomial $\tilde{r}_2(\varphi) = \frac{1}{2}r_2(\varphi) - \frac{1}{6}r_2'(\varphi - c) + \frac{1}{12}(\varphi - c)^2$ $= \frac{1}{4}\varphi^2 + \frac{4-3c}{6}\varphi + g + \frac{c(4-3c)}{12}$.
So

$$(\varphi')^2(\xi) = \frac{\bar{C}_1 + (\varphi - c)^2\tilde{r}_2(\varphi)}{(\varphi - c)^2} \Longleftrightarrow (\varphi - c)^2(\varphi')^2 = \bar{C}_1 + \tilde{r}_2(\varphi - c)^2, \varphi \neq c$$
$$(2.10)$$

and we must assume that $(\varphi - c)^2\tilde{r}_2(\varphi) + \bar{C}_1 \geq 0$ on some subinterval of $\mathbf{R}_\varphi^1$. Having in mind that Eq. (2.1)) is of the type (3.15) from [110] we shall apply our Theorems 3.1, 3.2 from [110] to Eq. (2.10). To find the solutions of Eq. (2.10) we shall consider two different cases (see [111]), namely

$(A1)$ $\quad \bar{C}_1 = 0 \Rightarrow (\varphi')^2 = \tilde{r}_2(\varphi)$
$(A2)$ $\quad \bar{C}_1 \neq 0.$

2.3 Investigation of the case (A1)

1. Denote by $\tilde{\Delta}(c, g)$ the discriminant of the quadratic equation $\tilde{r}_2(\varphi) = 0$. Then three different subcases appear:

$(A1)(a) : \tilde{\Delta} > 0$, i.e. $g < \frac{(c-1)^2}{2} - \frac{1}{18}$

$(A1)(b) : \tilde{\Delta} < 0$, i.e. $g > \frac{(c-1)^2}{2} - \frac{1}{18}$

$(A1)(c) : \tilde{\Delta} = 0$, where $\tilde{\Delta} = \frac{(c-1)^2}{2} - \frac{1}{18} - g$.

Put $X(\varphi) = a\varphi^2 + b\varphi + \bar{c}$, $a > 0$, $\Delta = b^2 - 4a\bar{c}$. Then according to [44]
$\int \frac{d\varphi}{\sqrt{X(\varphi)}} = \frac{1}{\sqrt{a}} log|2\sqrt{a}\sqrt{X(\varphi)} + 2a\varphi + b| = \frac{1}{\sqrt{a}} log|B|$.

We point out that in the case $(A1)(a) : \Delta > 0$, the equation $X(\varphi) = 0$ has two real distinct roots $\alpha < \beta$ and $B = 2\sqrt{a}\sqrt{X(\varphi)} + 2a\varphi + b > 0$ for $\beta < \varphi$, while $B < 0$ for $\varphi < \alpha$. In the case $(A1)(b) : \Delta < 0$ the function $B(\varphi) > 0$, $\forall \varphi$, while in the case $(A1)(c) : \Delta = 0 : \int \frac{d\varphi}{\sqrt{X(\varphi)}} =$

$\frac{1}{\sqrt{a}} \begin{cases} log(2a\varphi + b), & 2a\varphi + b > 0 \\ -log|2a\varphi + b|, & 2a\varphi + b < 0. \end{cases}$

Having in mind that in the case $(A1)(b) \int \frac{d\varphi}{\sqrt{X(\varphi)}} = \frac{1}{\sqrt{a}} Arsh \frac{2a\varphi + b}{\sqrt{4a\bar{c} - b^2}}$, $Arsh\ z$ being the inverse function of the strictly monotonically increasing function $sh\ w = \frac{e^w - e^{-w}}{2}$, $sh(-w) = -sh\ w$, $\frac{d}{dz} Arsh\ z = \frac{1}{\sqrt{1 + z^2}}$, we conclude that if $(A1)(b)$ holds, then $\pm \frac{d\varphi}{\tilde{r}_2(\varphi)} = \xi + \bar{C}_2$, $\bar{C}_2 = const \Rightarrow$

$$\varphi = \pm 2\sqrt{-\tilde{\Delta}} sh \frac{\xi + \bar{C}_2}{2} + c - \frac{4}{3}. \tag{2.11}$$

The solution φ is unbounded monotonically increasing (decreasing), $|\varphi(\pm\infty)| = \infty$.

Suppose now that $(A1)(c)$ holds. Then $(\varphi')^2 = \tilde{r}_2 = (\frac{\varphi}{2} + \frac{4-3c}{6})^2$ and after the standard translation $\varphi = \frac{3c-4}{3} + \eta$ we get $(\eta')^2 = \frac{1}{4}\eta^2 \Longleftrightarrow \eta' = \pm \frac{|\eta|}{2}$. Geometrically we have that if $\eta(0) = 0 \Rightarrow \eta \equiv 0$ and if $\eta(0) \neq 0$ then $\eta(\varphi)$ conserves its sign; $\eta = Ae^{-|\xi|/2}$ are peakons-solitons, while $\varphi = Ae^{-\frac{|\xi|}{2}} + c - 4/3$ satisfies Eq. (2.3).

Remark 2.1. Assume that $c = 4/3$. Then we obtain the Fornberg-Whitham solution of Eq. (2.1): $Ae^{-1/2|x - \frac{4}{3}t|}$.

We are going to study $(A1)(a)$ but we shall mention only that we have in this case an explicit formula for the unbounded solution too. In fact, if $\varphi > \beta$, $X(\alpha) = X(\beta) = 0$, we must solve with respect to φ the equation $Ke^{\sqrt{a}\xi} = \sqrt{4a^2\varphi^2 + 4ab\varphi + 4a\bar{c}} + 2a\varphi + b$, $a > 0$, $\varphi(0) = \beta$, $\varphi'(0) = 0$,

$\varphi(+\infty) = \infty, \beta = \frac{-b+\sqrt{b^2-4a\bar{c}}}{2a}, \alpha = \frac{-b-\sqrt{b^2-4a\bar{c}}}{2a}, b^2 - 4a\bar{c} > 0, K = const.$

Evidently, $\varphi = \frac{Ke^{\sqrt{a}\xi}+(b^2-4a\bar{c})e^{-\sqrt{a}\xi}\frac{1}{K}-2b}{4a}$ etc. Having in mind that $\xi = \int_\beta^\varphi \frac{d\lambda}{\sqrt{X(\lambda)}}, \varphi \geq \beta \iff \xi \geq 0$, we obtain: $\varphi = 2\sqrt{\tilde{\Delta}}ch\frac{\xi}{2} - 2b$, where $a = \frac{1}{4}$, $b = \frac{2}{3} - \frac{c}{2}, \bar{c} = g + \frac{(4-3c)c}{12}$.

Proposition 2.1. *([111])*

Consider the Fornberg-Whitham equation in the case (A1). Depending on the sign of the discriminant $\tilde{\Delta}$ there exist unbounded solutions $\varphi(\xi)$ that can be written into explicit form via elementary functions ((A1)(a), (A1)(b)). Peakon-soliton type solutions exist only in the case $\tilde{\Delta} = 0$. They can be written explicitly too.

2.4 Investigation of the case (A2)

1). Our first step is to do the translation $\varphi \to \varphi - c = \psi$ in Eq. (2.10) obtaining this way the ODE:

$$\psi^2(\psi')^2 = \bar{C}_1 + \psi^2\tilde{r}_2(c+\psi) \equiv P_4(\psi), Q_4(\psi) = \psi^2\tilde{r}_2(c+\psi), \qquad (2.12)$$

$P_4(\psi)$ being a 4th order polynomial, $Q_4(\psi) = \frac{1}{4}\psi^4 + \frac{2}{3}\psi^3 + (\frac{c}{2}(2-c)+g)\psi^2$; $\tilde{r}_2(\psi + c) = \frac{\psi^2}{4} + \frac{2}{3}\psi + L = (\frac{\psi}{2} + \frac{2}{3})^2 + L - \frac{4}{9}, L = c - \frac{c^2}{2} + g; L > \frac{4}{9} \Rightarrow$ $\tilde{r}_2(\psi + c) > 0, \forall\psi; L \leq 4/9$ implies that $\psi_{1,2}^0 = -\frac{4}{3} \pm 2\sqrt{4/9 - L}$ are real roots of $\tilde{r}_2(\psi + c) = 0$. We have double $-\frac{4}{3}$ root only for $L = \frac{4}{9}$.

The discriminant of $\tilde{r}_2(c + \psi)$ coincides with $\tilde{\Delta}$.

Evidently, $Q_4'(\psi) = 2\psi\tilde{r}_2(c+\psi) + \psi^2\tilde{r}_2'(c+\psi) = \psi(\psi^2 + 2\psi + c(2-c) + 2g) = \psi[(\psi+1)^2 - 1 + 2L]$. The discriminant of the quadratic polynomial in the brackets $\Delta_1 = 8(\frac{(c-1)^2}{2} - g)$ and therefore $\Delta_1 > 0 \iff g < \frac{(c-1)^2}{2}$; $\Delta_1 < 0 \iff g > \frac{(c-1)^2}{2}$. Therefore, $\frac{(c-1)^2}{12} - \frac{1}{18} < g < \frac{(c-1)^2}{2} \Rightarrow \tilde{\Delta} < 0$, $\Delta_1 > 0$, while $g > \frac{(c-1)^2}{2} > \frac{(c-1)^2}{12} - \frac{1}{18} \Rightarrow \Delta_1 < 0, \tilde{\Delta} < 0$.

Combining the above given remarks we come to the following figures, illustrating the geometrical structure of $Q_4(\psi)$ in each possible case : $L = 4/9 \Rightarrow \psi_{1,2}^0 = -4/3; 4/9 \leq L \Rightarrow \psi_1^0 \leq -4/3; 0 < L < \frac{4}{9} \Rightarrow \psi_1^0 < -4/3$, $-4/3 < \psi_2^0 < 0; L = 0 \Rightarrow \psi_1^0 = -\frac{8}{3}, \psi_2^0 = 0; L < 0 \Rightarrow \psi_2^0 > 0$ and $L \to -\infty \Rightarrow \psi_1^0 \to -\infty, \psi_2^0 \to +\infty, \psi_1^0 + \psi_2^0 = -8/3$. Put $\tilde{r}_2(\psi) = (\psi+1)^2 + 2L - 1; Q_4' = \psi\tilde{r}_2$. Then if $L > \frac{1}{2} \Rightarrow \tilde{r}_2(\psi) > 0, \forall\psi; L = 1/2 \Rightarrow \tilde{r}_2 = (\psi+1)^2; L < 1/2 \Rightarrow \tilde{r}_2(\psi) = 0$ has two real roots $\tilde{\psi}_{1,2} = -1\pm\sqrt{1 - 2L}$ such that $\tilde{\psi}_1 < -1; 0 < L < \frac{1}{2} \Rightarrow \tilde{\psi}_1 < \tilde{\psi}_2 < 0; L = 0 \Rightarrow \tilde{\psi}_1 = -2, \tilde{\psi}_2 = 0;$

$L < 0 \Rightarrow \tilde{\psi}_1 < 0,\ \tilde{\psi}_2 > 0,\ \tilde{\psi}_1 \to_{L \to -\infty} -\infty,\ \tilde{\psi}_2 \to_{L \to -\infty} +\infty,\ \dfrac{|\tilde{\psi}_2|}{|\tilde{\psi}_1|} < 1,$

$\dfrac{\tilde{\psi}_2}{\tilde{\psi}_1} \to_{L \to -\infty} -1.$ $\psi_{1,2}^0$ are the roots of $\tilde{r}_2(c + \psi) = 0$.

Having in mind that $Q_4 = \psi^2 \tilde{r}_2(\psi + c)$, $Q_4' = \psi \tilde{r}_2(\psi)$ we come to Fig. 2.1.

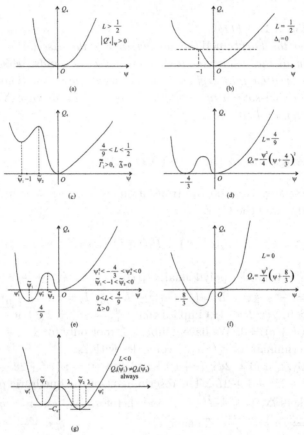

Fig. 2.1. Graph of Q_4.

We point out that in each subdomain $\{\psi \geq \varepsilon_0 < 0\}$ or $\{\psi \leq -\varepsilon_0 < 0\}$ the solution of Eq. (2.12) can be found easily as $\dfrac{P_4(\psi)}{\psi^2} \in C^\infty$ there (see [13]). So the most interesting case is the following one: Find the solution ψ of Eq. (2.12) for $0 \leq \psi$, respectively, $\psi \geq 0$. We shall concentrate now on a typical example, namely (e) from Fig. 2.1. Then $\tilde{r}_2(\psi + c) = \dfrac{\psi^2}{4} + \dfrac{2}{3}\psi + L$; $0 < L < \dfrac{4}{9}$ implies $\psi_{1,2}^0 = -\dfrac{4}{3} \pm 2\sqrt{4/9 - L}$. See now Fig. 2.2.

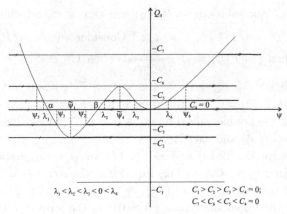

Fig. 2.2

The function $Q_4(\psi)$ has a local minimum at the point $\tilde{\psi}_1 \in (\alpha, \beta)$, where $Q_4(\alpha) = Q_4(\beta) = 0$ and for simplicity we put $\alpha = \psi_1^0$, $\beta = \psi_2^0$. Thus $Q_4'(\tilde{\psi}_1) = 0$. Put $-C_2 = Q_4(\tilde{\psi}_1) < 0$. Therefore, $0 > \psi > \tilde{\psi}_1 \Rightarrow Q_4(\psi) > -C_2 \Rightarrow P_4^{(1)} = C_2 + Q_4 > 0$, $P_4^{(1)}(\tilde{\psi}_1) = 0$, $P_4^{(1)}(0) = C_2 + Q_4(0) = C_2 > 0$, $(P_4^{(1)})'(\tilde{\psi}_1) = 0$. Assuming that $0 > -C_3 > -C_2 \iff 0 < C_3 < C_2$ we denote $-C_3 = Q_4(\psi_2)$, $\beta > \psi_2 > \psi_1$, $Q_4'(\psi_2) \neq 0$. Then $0 > \psi > \psi_2 \Rightarrow -C_3 < Q_4(\psi) \Rightarrow P_4^{(2)} = C_3 + Q_4(\psi) > 0$; $P_4^{(2)}(0) = C_3 > 0$, $P_4^{(2)}(\psi_2) = 0$, $(P_4^{(2)})'(\psi_2) \neq 0$. As we mentioned at the beginning we shall apply Theorems 3.1, 3.2 from [110] to:

$$\psi^2(\psi')^2 = P_4^{(1)}(\psi) \quad \text{for} \quad 0 > \psi > \tilde{\psi}_1, \tag{2.13}$$

$$\psi^2(\psi')^2 = P_4^{(2)}(\psi) \quad \text{for} \quad 0 > \psi > \psi_2. \tag{2.14}$$

For the Eq. (2.13) there exists a cuspon-soliton solution $\psi(\xi) = \psi(-\xi)$ such that ψ is strictly monotonically decreasing on the half line $\xi \geq 0$, $\psi(0) = 0$, $lim_{\xi \to \infty} \psi(\xi) = \tilde{\psi}_1$ and $\psi \sim d_1 \xi^{1/2}$, $\xi \to +0$, $d_1 < 0$. As it concerns Eq. (2.14) we put $\frac{T}{2} = -\int_0^{\psi_2} \frac{\psi d\psi}{\sqrt{P_4^{(2)}(\psi)}} < 0$. Then there exists a strictly monotonically decreasing solution ψ on $[\frac{T}{2}, 0]$ for which $\psi \sim d_2(\xi - \frac{T}{2})^{1/2}$, $\xi \to \frac{T}{2} + 0$, $d_2 = const < 0$, respectively $\psi \sim \psi_2 + d_3 \xi^2$, $\xi \to 0$, $d_3 = const > 0$. In a standard way we continue ψ on $[\frac{T}{2}, T]$ by the formula $\psi(\frac{T}{2} + \xi) = \psi(\frac{T}{2} - \xi)$, $\xi \in [0, \frac{T}{2}]$ and then periodically on $\mathbf{R}_\xi^1$. This way we have obtained a periodic cuspon type solution.

Evidently, $P_4^{(3)} = Q_4(\psi) + C_1 > 0$, $\forall \psi \in \mathbf{R}^1$. Then we must solve the equation $\psi^2(\psi')^2 = P_4^{(3)}(\psi) > 0$. As it is obvious Eq. (2.13), Eq. (2.14) do not have unique solutions as $P_4(\psi_0) = 0 \Rightarrow \psi \equiv \psi_0$ satisfies $\psi^2(\psi')^2 = P_4(\psi)$ and $\psi\psi' \pm \sqrt{P_i^{(2)}(\psi)}, i = 1, 2$. Consider $\psi\psi' = \sqrt{P_4^{(3)}(\psi)}$ with $\psi(0) = 1$. Then $\xi = F(\psi) = \int_1^\psi \frac{\lambda d\lambda}{\sqrt{P_4^{(3)}(\lambda)}}$. Then $1 \le \psi < \infty \Rightarrow F'(\psi) > 0$, $F'(1) > 0$, $F(\psi) \to_{\psi\to\infty} +\infty$; $0 < \psi < 1 \Rightarrow F'(\psi) > 0$ but $F(\psi) < 0$; $F(0) = A < 0$; $F'(0) = 0$, $F'(\psi) < 0$ for $0 > \psi$, $F(\psi) \to_{\psi\to-\infty} +\infty$.

Therefore, the double-valued function $\psi = F^{-1}(\xi)$ looks like a parabola with vertex at $(0, A)$ and axes $A\xi$.

We can solve Eq. (2.13) and Eq. (2.14) for $\psi < \tilde{\psi}_1$, respectively for $\psi < \psi_3$ or for $\psi > 0$. Concerning Eq. (2.13) the solution ψ is a cuspon with cusp at the origin, $\psi \approx c_0|\xi|^{1/2}$, $c_0 > 0$, $\xi \to 0$, $\psi(-\xi) = \psi(\xi)$ and is located in the upper half-plane $\{\psi > 0\}$; in the strip $0 > \psi > \tilde{\psi}_1$ it is soliton-cuspon with cusp at 0 and for $\psi < \tilde{\psi}_1$ there exists a peakon with peak at the vertical axes, $\psi(-\xi) = \psi(\xi)$, $lim_{\xi\to\pm\infty}\psi(\xi) = \tilde{\psi}_1$. Concerning Eq. (2.14) the solution is again cuspon with cusp at 0 and is located in $\{\psi > 0\}$. In the strip $0 > \psi > \psi_2$ one can construct a periodic cuspon with period T such that the cusps are $(\xi_n = nT, 0)$, $\forall n \in \mathbf{Z}$. There are no solutions in the strip $\psi_3 < \psi < \psi_2$ while the solution ψ for $\psi \le \psi_3$ is a parabola with vertex at $(0, \psi_3)$ and axes $\psi_3\psi$, $\psi(-\xi) = \psi(\xi)$.

We shall continue our study on the level of the case proposed on Fig. 2.2; $P_4^{(5)} = Q_4 + C_5$, the case $C_4 = 0$ coincides with (A1). Then $P_4^{(5)}(\lambda_i) = 0$, $1 \le i \le 4$.

Evidently, Q_4 has a local maximum at $\tilde{\psi}_4 : Q_4(\tilde{\psi}_4) = -C_6, Q_4'(\tilde{\psi}_4) = 0$. Then $P_4^{(4)} = Q_4 + C_6$ is such that $P_4^{(4)}(\tilde{\psi}_4) = 0$, $(P_4^{(4)})'(\tilde{\psi}_4) = 0$; $P_4^{(4)}(\psi_5) = 0$, $(P_4^{(4)})'(\psi_5) \ne 0$ and $P_4^{(4)}(\psi_6) = 0$, $(P_4^{(4)})'(\psi_6) \ne 0$; $\psi_5 < \lambda_1 < \lambda_2 < \tilde{\psi}_4 < \lambda_3 < 0 < \lambda_4 < \psi_6$ and $\lambda_i, 1 \le i \le 4$ are the simple roots of the algebraic 4th order equation $P_4^{(5)}(\psi) = Q_4(\psi) + C_5 = 0$. From Fig. 2.2 it is obvious that $\psi \ne \tilde{\psi}_4$, $\psi_5 < \psi < \psi_6 \Rightarrow P_4^{(4)}(\psi) < 0$, while $\psi < \psi_5$ or $\psi > \psi_6 \Rightarrow P_4^{(4)}(\psi) > 0$. In a similar way $\psi < \lambda_1$, $\psi > \lambda_4$, $\lambda_2 < \psi < \lambda_3 \Rightarrow P_4^{(5)}(\psi) > 0$, while $\lambda_1 < \psi < \lambda_2$ or $\lambda_3 < \psi < \lambda_4 \Rightarrow P_4^{(5)}(\psi) < 0$.

Consequently, real valued nontrivial solutions of

$$\psi^2(\psi')^2 = P_4^{(4)}(\psi) \tag{2.15}$$

$$\psi^2(\psi')^2 = P_4^{(5)}(\psi) \tag{2.16}$$

exist for Eq. (2.15) in the cases $\psi \equiv \tilde{\psi}_4$, $\psi < \psi_5$ or $\psi > \psi_6$ and for Eq. (2.16) in the cases $\psi < \lambda_1$, $\psi > \lambda_4$ and $\lambda_2 < \psi < \lambda_3$. Equation (2.15) does not possess bounded nontrivial solutions, while there exist for Eq. (2.16) iff $\lambda_2 < \psi < \lambda_3 \Rightarrow P_4^{(5)}(\psi) > 0$.

Remark 2.2. $P_4^{(4)}(\psi) > 0$ $(P_4^{(5)}(\psi) > 0)$ on such intervals γ on the real line that $0 \notin \bar{\gamma}$. In other words then Eq. (2.15), Eq. (2.16) have the form: $(\psi')^2 = \frac{P_4^{(4)}(\psi)}{\psi^2} \in C^\infty$, respectively $(\psi')^2 = \frac{P_4^{(5)}(\psi)}{\psi^2} \in C^\infty$. Those cases are studied in details in [13] etc. As we know,

$$P_4^{(5)}(\psi) = \frac{1}{4}(\psi - \lambda_1)(\psi - \lambda_2)(\psi - \lambda_3)(\psi - \lambda_4), \qquad (2.17)$$

$\lambda_1 < \lambda_2 < \lambda_3 < 0 < \lambda_4$ and

$$P_4^{(4)}(\psi) = 1/4(\psi - \tilde{\psi}_4)^2(\psi - \psi_6)(\psi - \psi_5), \psi_5 < \tilde{\psi}_4 < 0 < \psi_6. \qquad (2.18)$$

2). In order to study Eq. (2.16) we must integrate the equation

$$\psi\psi' = \pm\sqrt{P_4^{(5)}(\psi)}, 0 < \lambda_2 \leq \psi < \lambda_3,$$

i.e.

$$\pm(\xi + M) = \int \frac{\psi d\psi}{\sqrt{P_4^{(5)}(\psi)}}, M = const. \qquad (2.19)$$

We know that in this case there exists a smooth periodic solution [13].

In [55] two formulas are proposed for the primitive of the right-hand side of Eq. (2.19). We give here only one of them, namely for $\lambda_2 < \psi < \lambda_3$:

$$I_1 = \int_\psi^{\lambda_3} \frac{2x dx}{\sqrt{(x - \lambda_1)(x - \lambda_2)(x - \lambda_3)(x - \lambda_4)}} = \qquad (2.20)$$

$$\frac{4}{\sqrt{(\lambda_4 - \lambda_2)(\lambda_3 - \lambda_1)}} \left\{ (\lambda_3 - \lambda_4)\Pi\left(\kappa, \frac{\lambda_3 - \lambda_2}{\lambda_4 - \lambda_2}, q\right) + \lambda_4 F(\kappa, q) \right\},$$

where $\kappa = arcsin\sqrt{\frac{(\lambda_4 - \lambda_2)(\lambda_3 - \psi)}{(\lambda_3 - \lambda_2)(\lambda_4 - \psi)}}$, $q = arcsin\sqrt{\frac{(\lambda_3 - \lambda_2)(\lambda_4 - \lambda_1)}{(\lambda_4 - \lambda_2)(\lambda_3 - \lambda_1)}}$ and Π, F are the famous Legendre's elliptic functions of third and first kind (see [8], [30], [55]).

As it is well known the solution we have constructed above is periodic and smooth everywhere. Its period $T = 4\int_{\lambda_2}^{\lambda_3} \frac{x dx}{\sqrt{(x - \lambda_1)(x - \lambda_2)(x - \lambda_3)(x - \lambda_4)}}$.

To complete the things we must solve the ODE Eq. (2.15), i.e. we must evaluate the integral $\frac{1}{2}I_2 = \int \frac{\psi d\psi}{(\psi - \tilde{\psi}_4)\sqrt{(\psi - \psi_5)(\psi - \psi_6)}}$, say for $\psi > \psi_6$. As the computations are standard via Euler's substitutions we omit them.

On the other hand, we can apply to Eq. (2.15) and for $\psi > \psi_6$ Theorem 3.2 from [110]. Details are left to the reader.

Proposition 2.2.

Considering the case (A2) we have found classical (smooth) periodic solutions of Eq. (2.12), "parabola" type solutions as well as soliton-cuspon, soliton-peakon and periodic cusp wave solutions.

"Parabola" type solutions exist for $\psi \leq \psi_5$, respectively for $\psi \geq \psi_6$. The corresponding parabolas have vertices at the lines $\psi = \psi_5$ ($\psi = \psi_6$). The same result holds for $\lambda \leq \lambda_1$ and $\lambda \geq \lambda_4$. The solution $\psi \equiv \tilde{\psi}_4$ is isolated in the strip $\psi_5 < \psi < \psi_6$ (i.e. there are no other solutions in that strip).

It is interesting to know whether for some value of the parameter $L < 0$: $Q_4(\tilde{\psi}_1) = Q_4(\tilde{\psi}_2)$. If so, we can construct a kink type solution of $\psi^2(\psi')^2 = Q_4 + C_2$, where $Q_4(\tilde{\psi}_1) = Q_4(\tilde{\psi}_2) = -C_2 < 0$. Then $lim_{\xi \to +\infty}\psi(\xi) = A$, $lim_{\xi \to -\infty}\psi(\xi) = B$, $A \neq B$. Certainly, $\psi(\xi)$ will have a vertical tangent at the point $(\xi_0, \psi(\xi_0) = 0)$. To solve

$$Q_4(\tilde{\psi}_1) = Q_4(\tilde{\psi}_2) \tag{2.21}$$

we put $u = \sqrt{1 - 2L} > 1$ for $L < 0$, i.e. $\tilde{\psi}_1 = -1 - u$, $\tilde{\psi}_2 = -1 + u$, $L = \frac{1-u^2}{2} < 0$, $L - \frac{4}{9} = \frac{1}{18} - \frac{u^2}{2}$, $Q_4(\psi) = \psi^2[(\frac{\psi}{2} + \frac{2}{3})^2 + L - \frac{4}{9}]$. One can easily see that Eq. (2.21) is a cubic equation with respect to u but unfortunately it does not possess any root $u > 1$. In fact, it has triple root $u = 0$. Therefore, the configuration on Fig. 2.1(g) does not give rise to a kink having one vertical tangent only.

Below we propose a new and interesting case when the configuration loop of the solutions appears. Consider again Fig. 2.1(g) $\Rightarrow$ without loss of generality $Q_4(\tilde{\psi}_1) < Q_4(\tilde{\psi}_2)$. Let $Q_4(\tilde{\psi}_1) = Q_4(\lambda_1) = -C_0 = Q_4(\lambda_2) < 0$ and put $P_4^{(6)} = Q_4 + C_0$. Then $P_4^{(6)}(\tilde{\psi}_1) = P_4^{(6)}(\lambda_1) = 0$, $\tilde{\psi}_1 < \psi < \lambda_1 \Rightarrow$ $P_4^{(6)}(\psi) > 0$, $(P_4^{(6)})'(\tilde{\psi}_1) = 0$, $(P_4^{(6)})'(\lambda_1) < 0$, $P_4^{(6)}(\lambda_2) = 0$, $(P_4^{(6)})'(\lambda_2) > 0$. Therefore, $\tilde{\psi}_1$ is a double root and λ_1, λ_2 are simple roots of the 4th order algebraic equation $P_4^{(6)}(\psi) = 0 \Rightarrow (P_4^{(6)})''(\tilde{\psi}_1) > 0$, $P_4^{(6)}(0) = C_0 > 0$. We shall rely heavily on the non-uniqueness result. At first we construct soliton-cuspon solution $\bar{\psi}$ of $\psi^2(\psi')^2 = P_4^{(6)}(\psi)$, $\bar{\psi}(0) = 0$, such that $\bar{\psi}(-\xi) = \bar{\psi}(\xi)$ and $\bar{\xi}$ has a cusp at 0, asymptote $\psi = \tilde{\psi}_1$ at $\xi = \pm\infty$ and $\bar{\psi}(\xi)$ is located in the strip $\{\tilde{\psi}_1 < \bar{\psi} \leq 0\}$. Define $0 < \frac{T_1}{2} = \int_0^{\lambda_1} \frac{\gamma d\gamma}{\sqrt{P_4^{(6)}(\gamma)}}$ and construct a cuspon solution $\bar{\bar{\psi}}$, $\bar{\bar{\psi}}(0) = 0$, with cusp at 0 and located in $\{0 \leq \bar{\bar{\psi}} \leq \lambda_1\}$,

$\bar{\bar{\psi}}(-\xi) = \bar{\bar{\psi}}(\xi)$ and $\bar{\bar{\psi}}(\frac{T_1}{2}) = \lambda_1$, $\bar{\bar{\psi}}'(\frac{T_1}{2}) = 0$. Combining $\bar{\psi}$, $\bar{\bar{\psi}}$ we get double valued left branch of $\psi = \begin{cases} \bar{\psi} \\ \bar{\bar{\psi}} \end{cases}$ defined for $\xi \leq 0$ and right branch defined for $\xi \geq 0$. The standard translations $\xi \to \xi \pm \frac{T_1}{2}$ of these branches give us the triple-valued configuration loop Γ (see Fig. 2.3) where Γ is analytic (locally) outside $(\pm\frac{T_1}{2}, 0)$.

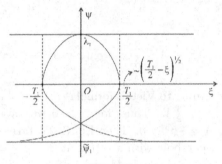

Fig. 2.3

2.5 Traveling wave solutions of the Vakhnenko equation. Geometrical interpretation

To find traveling wave solution of Eq. (2.2) we look for $u = \varphi(x - ct)$, $\xi = x - ct$, $c = const$. Thus,

$$\varphi''(\varphi - c) + (\varphi')^2 + \varphi = 0. \tag{2.22}$$

$\varphi \equiv c$ satisfies Eq. (2.22) if and only if $c = 0$. The standard change $\varphi'(\xi) = p(\varphi) \Rightarrow \varphi'' = p\frac{dp}{d\varphi}$ leads to

$$\frac{d}{d\varphi}(p^2)(\varphi - c) + 2p^2 + 2\varphi = 0,$$

i.e. the second change $q = p^2(\varphi)$ implies

$$(\varphi - c)\frac{dq}{d\varphi} + 2q + 2\varphi = 0. \tag{2.23}$$

As Eq. (2.23) is linear ODE in q, $\frac{dq}{d\varphi}$ we get

$$q = \frac{1}{(\varphi - c)^2}\left(A - 2\left(\frac{\varphi^3}{3} - \frac{c\varphi^2}{2}\right)\right), \tag{2.24}$$

A being arbitrary constant. Thus

$$q = p^2 = (\varphi')^2 = \frac{A - \frac{2}{3}\varphi^3 + c\varphi^2}{(\varphi - c)^2}. \qquad (2.25)$$

Therefore, we shall assume that $P_3(\varphi) = -\frac{2}{3}\varphi^3 + c\varphi^2 + A \geq 0$.

We consider the following two cases:

$$(A3): \quad -\frac{2}{3}\varphi^3 + c\varphi^2 + A \equiv \frac{2}{3}(\varphi - \alpha)^2(\beta - \varphi), \left| \begin{matrix} \alpha < \beta \text{ (double root)} \\ \alpha < \varphi < \beta, \alpha\beta \neq 0 \end{matrix} \right.$$

$$(A4): \quad -\frac{2}{3}\varphi^3 + c\varphi^2 + A = \frac{2}{3}(\varphi - \alpha)(\varphi - \beta)(-\varphi + \gamma), \left| \begin{matrix} \alpha < \beta < \gamma \\ \beta < \varphi < \gamma \\ \alpha\beta\gamma \neq 0. \end{matrix} \right.$$

In the case (A3) according to Viete's formulas $\beta + 2\alpha = \frac{3}{2}c$, $\alpha(\alpha + 2\beta) = 0$, i.e. $\alpha = -2\beta$, $\alpha^2\beta = \frac{3}{2}A$. Thus, for $\beta > 0$ we have $\alpha = -2\beta < 0$, $c = -2\beta = \alpha < 0$, $A = \frac{8}{3}\beta^3 > 0$. Then Eq. (2.25) takes the form $(\varphi')^2 = \frac{2}{3}(\beta - \varphi)$, i.e. φ is a parabola with a vertex at $\beta = \varphi$.

Case (A4) is more complicated. Then $\alpha + \beta + \gamma = \frac{3}{2}c$, $\alpha(\beta + \gamma) + \beta\gamma = 0$, $\alpha\beta\gamma = \frac{3}{2}A$. Therefore, if $0 < \beta < \gamma$ then $\alpha = -\frac{\beta\gamma}{\beta+\gamma} < 0$, $\alpha = \frac{3A}{2\beta\gamma}$, $\alpha = \frac{3}{2}c - (\beta + \gamma)$.

The overdetermined system for α with $0 < \beta < \gamma$ parameters is solvable if $A = -\frac{2}{3}\frac{\beta^2\gamma^2}{\beta+\gamma} < 0$ and $\frac{3}{2}c(\beta+\gamma) - (\beta+\gamma)^2 = -\beta\gamma \iff c = \frac{2}{3}\frac{\beta^2+\gamma^2+\beta\gamma}{\beta+\gamma} > 0$.

As $0 < \beta < \gamma$ are arbitrary, we are interested for which choice of those parameters

$$0 < \beta < c < \gamma \qquad (2.26)$$

$$\left(\iff \beta^2 + \frac{\beta\gamma}{3} < \frac{2}{3}(\beta^2 + \gamma^2) < \frac{\beta\gamma}{3} + \gamma^2 \right.$$

$$\iff \frac{\beta}{\gamma}\left(\frac{\beta}{\gamma} + \frac{1}{3}\right) < \frac{2}{3}\left(\frac{\beta^2}{\gamma^2} + 1\right) < 1 + \frac{1}{3}\frac{\beta}{\gamma} \right).$$

Taking $0 < \beta = o(\gamma)$, $\gamma \to \infty$ we verify Eq. (2.26).

We shall solve in that case II: $\alpha < 0 < \beta < c < \gamma$,

$$\left| \begin{matrix} \varphi' = \sqrt{P_3(\varphi)}\frac{1}{|\varphi-c|} \\ \varphi(0) = \varphi_0 \in (\beta, c), \end{matrix} \right. \qquad (2.27)$$

i.e.

$$\xi = \sqrt{\frac{3}{2}} \int_{\varphi_0}^{\varphi} \frac{|\lambda - c|}{\sqrt{(\lambda - \alpha)(\lambda - \beta)(\gamma - \lambda)}} d\lambda = F(\varphi), \beta < \varphi < \gamma. \qquad (2.28)$$

Evidently, $F'(\varphi) > 0$ for $\varphi \neq c$, $F'(c) = 0$, $F'(\beta) = \infty$, $F'(\gamma) = \infty$, $F(\varphi_0) = 0$, the integral is convergent at $\varphi = \beta$, $\varphi = \gamma$, $\beta_1 = F(\beta) = \sqrt{\frac{3}{2}} \int_{\varphi_0}^{\beta} \frac{|\lambda - c|}{\sqrt{(\lambda - \alpha)(\lambda - \beta)(\gamma - \lambda)}} d\lambda < 0$, $F(\gamma) = \gamma_1 > 0$, $F(c) = c_1 > 0$.

The inverse function $\varphi = F^{-1}(\xi)$ is strictly monotonically increasing in $[\beta_1, \gamma_1]$, $\varphi(\beta_1) = \beta$, $\varphi'(\beta_1) = 0$, $\varphi(0) = \varphi_0$, $\varphi(c_1) = c$, $\varphi'(c_1) = \infty$, $\varphi(\gamma_1) = \gamma$, $\varphi'(\gamma_1) = 0$ and $\varphi'(\xi) < \infty$ for $\xi \neq c_1$. Put $\frac{T}{2} = \gamma_1 - \beta_1 = \sqrt{\frac{3}{2}} \int_{\beta}^{\gamma} \frac{|\lambda - c|}{\sqrt{(\lambda - \alpha)(\lambda - \beta)(\gamma - \lambda)}} d\lambda > 0$.

We continue φ in an even way with respect to γ_1, i.e. $\varphi(\gamma_1 + \tau) = \varphi(\gamma_1 - \tau)$ for $0 \leq \tau \leq \frac{T}{2}$ and then periodically: $\varphi(\xi + T) \equiv \varphi(\xi)$. Below we propose Fig. 2.4 for illustration of that result:

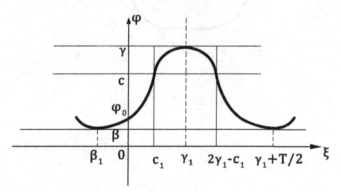

Fig. 2.4

The curve at Fig. 2.4 is single valued, periodic and possesses vertical tangents at countably many points ($\xi = c_1$, $\xi = 2\gamma_1 - c_1$, forming arithmetic progression).

There are other possibilities for continuing $\varphi(\xi)$ as multivalued function forming the configurations loop and butterfly. In fact, take the branch of $\varphi(\xi)$, $\gamma_1 < \xi < 2\gamma_1 - c_1$, $(\varphi')^2 = \frac{P_3(\varphi)}{(\varphi - c)^2}$. The translation $\eta = \xi + 2(c_1 - \gamma_1)$ transforms $(\gamma_1, 2\gamma_1 - c_1)$ into $(-\gamma_1 + 2c_1, c_1)$, $-\gamma_1 + 2c_1 < c_1$. Certainly, $\varphi(\eta)$ satisfies the same equation for $\eta \in (-\gamma_1 + 2c_1, c_1)$. The double valued curve $\varphi(\xi)$ defined by Eq. (2.28) on $[\beta_1, c_1]$ and by $\varphi(\eta)$ on $(-\gamma_1 + 2c_1, c_1)$ can be continued in an even way with respect to $2c_1 - \gamma_1 = T_1 < c_1$. This way we get the following Fig. 2.5 and Fig. 2.6:

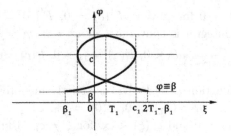

Fig. 2.5. $T_1 > \beta_1$-loop.

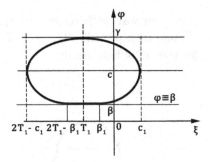

Fig. 2.6. $T_1 < \beta_1$-oval.

We shall complete those considerations with the configuration butterfly (see Fig. 2.7).

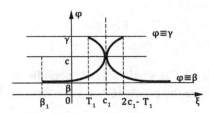

Fig. 2.7. Butterfly.

The branches of φ are symmetric with respect to the line $\xi = c_1$.

Remark 2.3. The integral $I = \int_\beta^\varphi \frac{|\varphi - c|d\varphi}{\sqrt{(\varphi-\alpha)(\varphi-\beta)(\gamma-\varphi)}} = \int_\beta^\varphi \frac{(c-\varphi)d\varphi}{\sqrt{(\varphi-\alpha)(\varphi-\beta)(\gamma-\varphi)}}$ for $\varphi \leq c$, while $I = \int_\beta^\varphi \frac{(\varphi-c)d\varphi}{\sqrt{(\varphi-\alpha)(\varphi-\beta)(\gamma-\varphi)}} + 2\int_\beta^c \frac{(c-\varphi)d\varphi}{\sqrt{(\varphi-\alpha)(\varphi-\beta)(\gamma-\varphi)}}$.

On the other hand,

$$I_1(\varphi) = \int_\beta^\varphi \frac{(\varphi - c)d\varphi}{\sqrt{(\varphi - \alpha)(\varphi - \beta)(\gamma - \varphi)}} =$$

$$I_1(\gamma) + \int_\gamma^\varphi \frac{(\varphi - c)d\varphi}{\sqrt{(\varphi - \alpha)(\varphi - \beta)(\gamma - \varphi)}}.$$

It is well known that [30], [55] $\int_\varphi^\gamma \frac{\varphi d\varphi}{\sqrt{(\varphi-\alpha)(\varphi-\beta)(\gamma-\varphi)}} = 2\frac{\alpha F(z,k)}{\sqrt{\gamma-\alpha}} +$ $2\sqrt{\gamma - \beta}E(z,k)$, where $k = \sqrt{\frac{\gamma-\beta}{\gamma-\alpha}}$, $z = arcsin\sqrt{\frac{\gamma-\varphi}{\gamma-\beta}}$ and $F(z,k)$, $E(z,k)$ are incomplete elliptic integrals of the first and second kind (Legendre functions). The single loop and the butterfly can be considered as multi-valued soliton solutions or as double (triple) valued compactons. One can find also solutions of Eq. (2.27) comprising periodic loops.

Remark 2.4. It is interesting to point out that Eq. (2.22) does not possess a classical solution $\varphi \in C^2(\mathbf{R}^1)$, such that $lim_{|x|\to\infty}\varphi(x) = lim_{|x|\to\infty}\varphi'(x) = 0$ and $\varphi(x) \neq 0$ for $|x| > A$, $A = const$.

In fact, if such φ exists define the function $v(x) = +c\varphi - \frac{\varphi^2}{2} \Rightarrow v'' = \varphi$. Multiplying $v(x)$ by φ' and integrating over the interval $(-\infty, x)$ we get:

$$\int_{-\infty}^x v(\lambda)\varphi'(\lambda)d\lambda = \frac{c}{2}\varphi^2 - \frac{1}{6}\varphi^3 \Rightarrow \frac{c}{2}\varphi^2 - \frac{1}{6}\varphi^3 = \varphi v - \frac{1}{2}(v')^2.$$

Integrating Eq. (2.22) over $\mathbf{R}^1$ we obtain $\int_{-\infty}^\infty \varphi(x)dx = 0$. Denote by ξ_1 the smallest zero of φ, i.e. $\varphi \neq 0$ for $x \in (-\infty, \xi_1)$, $\varphi(\xi_1) = 0$. Evidently, $v'(\xi_1) = 0$. Having in mind that $v'(x) = \int_{-\infty}^x \varphi(\lambda)d\lambda$ we see that $v' \neq 0$ on the ray $(-\infty, \xi_1)$, $lim_{x\to-\infty}v'(x) = 0$.

Therefore, $v'(x)$ attains its maximum (minimum) at the point $\xi_2 < \xi_1$, $v'(\xi_2) \neq 0 \Rightarrow v''(\xi_2) = 0 = \varphi(\xi_2)$. The contradiction proves our Remark.

To complete this Chapter we shall consider the 2D generalized Benney-Luke Eq. (2.29) [91]:

$$u_{tt} - \Delta_2 u + a\Delta_2^2 u - b\Delta_2 u_{tt} + pu_t(u_x^{p-1}u_{xx} + u_y^{p-1}u_{yy}) + \qquad (2.29)$$

$$2(u_x^p u_{xt} + u_y^p u_{yt}) = 0, p \in \mathbf{N}.$$

The original Benney-Luke 1D equation is obtained for $u = u(t,x)$, $p = 1$. Certainly, $\Delta_2 = \frac{\partial^2}{\partial x^2} + \frac{\partial^2}{\partial y^2}$. Put $u = \varphi(x + \beta y - ct)$, $\xi = x + \beta y - ct$, $b = const$, $c = const$. Then Eq. (2.29) implies

$$(c^2 - 1 - \beta^2)\varphi'' + (1+\beta^2)(a - bc^2 + a\beta^2)\varphi'''' - \varphi'^p\varphi''(1+\beta^{p+1})c(p+2) = 0.$$

$$(2.30)$$

Let $(a - bc^2 + a\beta^2) \neq 0$. Put $A = \frac{1+\beta^2-c^2}{(1+\beta^2)(a-bc^2+a\beta^2)} \neq 0$ and $B = \frac{c(1+\beta^{p+1})(p+2)}{(p+1)(1+\beta^2)(a-bc^2+a\beta^4)}$. Integrating Eq. (2.30) with respect to ξ we get

$$(c^2 - 1 - \beta^2)\varphi' + (1 + \beta^2)(a - bc^2 + a\beta^2)\varphi''' -$$

$$\frac{c(p+2)(1+\beta^{p+1})}{p+1}\varphi'^{p+1} = \tilde{C} = const.,$$

i.e. the change $\varphi' = \psi(\xi)$ leads to $\psi'' = A\psi + B\psi^{p+1} + \tilde{\tilde{C}}$, $\tilde{\tilde{C}} = const.$, i.e.

$$(\psi')^2 = A\psi^2 + 2B\frac{\psi^{p+2}}{p+2} + 2\tilde{\tilde{C}}\psi + \tilde{\tilde{\tilde{C}}}, \tilde{\tilde{\tilde{C}}} = const. \qquad (2.31)$$

Can you propose a full classification of the traveling wave solutions of Eq. (2.29)? In the case $p = 1, 2$ the solutions of Eq. (2.31) can be expressed via elliptic functions.

Can you find explicit formulas for the solutions $\varphi = \int \psi d\xi$ starting from the solutions of Eq. (2.31)? What about the existence of periodic, kink and soliton type solutions of Eq. (2.30)?

Hint: As we know for $p = 1$ (see [44], [110]) the Eq. (2.31) can have both periodic and soliton type solutions when the polynomial in the right-hand side of Eq. (2.31) possesses real roots only. These solutions are expressed via $sech^2 D\xi$, respectively $cn^2 D\xi$, $D = const.$ Therefore, $\varphi(\xi)$ takes the form $C_1\xi + C_2 thD\xi$, respectively $C_1 + C_2\{E(am\ D\xi, k) - (k')^2 D\xi\}$, where $(k')^2 + k^2 = 1$, $0 < k^2 < 1$, $E(\lambda, k)$ is elliptic integral of second kind (Legendre function) and $am\ u$ is the inverse function of $u = \int_0^\varphi \frac{d\varphi}{\sqrt{1-k^2sin^2\varphi}}$, i.e. $u = \int_0^{am\ u} \frac{d\varphi}{\sqrt{1-k^2sin^2\varphi}}$ (see [44]). Put in Eq. (2.31) $A > 0$, $\frac{2B}{p+2} = D < 0$, $p = 2$, $\tilde{\tilde{C}} = 0 = \tilde{\tilde{\tilde{C}}}$. Then $\psi(\xi) = \sqrt{\frac{-A}{D}}sech(\xi - x_0)\sqrt{A} \Rightarrow \varphi = \sqrt{\frac{-1}{D}}arctg(sh(\xi-x_0)\sqrt{A}) \Rightarrow \varphi$ is kink. Let $A = 2k^2 - 1$, $B = -2k^2$, $p = 4$, $\tilde{\tilde{C}} = 2k^2 - 1$. Then $\psi = cn(\xi, k)$, $0 < k^2 < 1$ is a solution of Eq. (2.31), i.e. $\varphi(\xi) = \frac{1}{k}arccos(dn(\xi, k))$ is a periodic solution of Eq. (2.30) with period $2K(k)$ [44], [55].

Chapter 3

Explicit formulas to the solutions of several equations of physics and geometry

3.1 Introduction

This chapter deals initially with the semilinear multidimensional Klein-Gordon equation, the wave equation, Kadomtsev-Petviashvilli equation and cubic first order hyperbolic pseudo diffential equation (Szegö equation). All these equations have interesting applications in different areas of Mathematical Physics and technics (see [121], [138]). Concerning Klein-Gordon (K-G), respectively multidimensional wave equations, we can construct special solutions via appropriate change of the unknown function u and by solving some overdetermined systems of linear and nonlinear PDE. We recall that in the case of 1D sin-Gordon equation solutions containing elliptic and hyperbolic functions appear [110]. The wave solutions of Kadomtsev-Petviashvili (K-P) equation are constructed via Hirota method [5], [63], [122] but their interaction can give rise to the so-called X and Y waves. In studying the linear first order multidimensional wave equation (periodic case in (t, x)) the machinery of small denominators works [14], while in the cubic case we are able to construct classes of Szegö type solutions [50], [51].

At the second part of this chapter we investigate Tzitzeika equation [128], [129] with solution v. Since that equation is not popular we shall mention that it appears in classical differential geometry as compatibility condition of 3 linear second order partial differential equations satisfied by the function $\theta(x, y)$. If the above mentioned system possesses 3 linearly independent solutions Θ^1, Θ^2, Θ^3 then the equations $x^i = \Theta^i(x, y)$, $i = 1, 2, 3$ define in $\mathbf{R}^3$ the parametric representation of a surface T. v is called a potential of T for v, v_x, v_y participate in that system. Our aim here is to construct exact solutions of Tzitzeika equation via different approaches.

Explicit formulas to the solutions of other classes of nonlinear systems of PDE are found via geometrical methods or methods coming from ODE in [1], [43], [54], [107]. Because of the lack of space we do not provide here the corresponding examples. We recommend to the reader the above mentioned papers and books for more details.

3.2 Special solutions of semilinear K-G type equations in the multidimensional case

Consider the multidimensional K-G PDE

$$Lu + (\nabla_{t,x}u, \vec{B}) = f(u) \quad \text{in } \mathbf{R}_t^1 \times \mathbf{R}_x^n, n \geq 2, \tag{3.1}$$

where $L = \partial_t^2 - \sum_{j=1}^n \frac{\partial^2}{\partial x_j^2}$ is the wave operator, the constant vector $\vec{B} = (b_0, b_1, \ldots, b_n) \in \mathbf{R}_t^1 \times \mathbf{R}_x^n$, the real-valued function $f \in C^1(\mathbf{R}^1)$.

According to the classical approach (see [7]), we look for a solution of Eq. (3.1) of the form $u = \varphi(G)$, $\varphi \in C^2(\mathbf{R}^1)$, $G = G(t, x) \in C^2$.

Then Eq. (3.1) takes the form

$$\varphi' LG + \varphi'' \left(G_t^2 - \sum_{j=1}^n G_{x_j}^2 \right) + \varphi' (\nabla_{t,x}G, \vec{B}) = f(\varphi(G)). \tag{3.2}$$

Further on we shall assume that either

$$\left| \begin{array}{l} LG + (\nabla_{t,x}G, \vec{B}) = -G \\ \sum_{j=1}^n G_{x_j}^2 - G_t^2 = G^2, \end{array} \right. \tag{3.3}$$

i.e.

$$G\varphi'(G) + G^2\varphi''(G) = -f(\varphi(G)) \tag{3.4}$$

or

$$\left| \begin{array}{l} LG + (\nabla_{t,x}G, \vec{B}) = 0 \\ \sum_{j=1}^n G_{x_j}^2 - G_t^2 = 1, \end{array} \right. \tag{3.5}$$

i.e.

$$\varphi''(G) = -f(\varphi(G)) \quad \text{(pendulum equation)}. \tag{3.6}$$

The change $G = e^t$ in the Euler Eq. (3.4) leads to $\tilde{\varphi}''(t) = -f(\tilde{\varphi}(t))$. Solving the pendulum equation we get $\tilde{\varphi} = \tilde{\varphi}(t) \Rightarrow \varphi = \tilde{\varphi}(\ln G), G > 0$.

Example 1. We shall illustrate Eq. (3.1) with the following examples:
a). $f(u) = e^{-2u}$, b). $f(u) = -u(\ln u + \ln^2 u)$, c). $f(u) = -\text{sh } u$, d). $f(u) = \pm\sin u$, e). $f(u) = -\frac{1}{2}e^u$, f). $f(u) = 3u^2 - 3\beta^2$, $\beta < 0$.

In case a). we take $u = \varphi(G) = \ln G, > 0$ which satisfies Eq. (3.6); in case b). we take $u = \varphi(G) = e^G$, which satisfies Eq. (3.4); in case c). Eq. (3.6) takes the form $\varphi'' = sh\,\varphi \Rightarrow (\varphi')^2 = 2(ch\,\varphi + A)$ and therefore we take $\varphi = 2\ln|tg\frac{G}{2}|$ for $A = 1$, $\varphi = 2\ln|cth\frac{G}{2}|$ for $A = -1$; in case e). the Eq. (3.6) is written as $\varphi'' = \frac{1}{2}e^\varphi \Rightarrow \varphi' = \pm e^{\frac{\varphi}{2}}$ and we put $u = \varphi(G) = -2\ln|\frac{G}{2}|$, $G \neq 0$; in case d). $f(u) = -\sin u$ the Eq. (3.4) possesses the solution $\varphi = 4arctg\,G$, while if $f(u) = \sin u$ we take $\varphi = 4(arctgG - \frac{\pi}{4})$. In case f). and after the change $G = e^t$ the Eq. (3.4) can be written as: $\tilde{\varphi}'' = -(3\tilde{\varphi}^2 - 3\beta^2) \Rightarrow$

$$(\tilde{\varphi}')^2 = -2(\tilde{\varphi}^3 - 3\beta^2\tilde{\varphi} + 2\beta^3), \qquad (3.7)$$

the constant $2\beta^3$ being appropriate chosen after the integration. As $\tilde{\varphi} = \beta$ is a double root of $\tilde{\varphi}^3 - 3\beta^2\tilde{\varphi} + 2\beta^3 = 0$ and $\tilde{\varphi} = -2\beta > 0$ is a simple root we can integrate Eq. (3.7) obtaining $\varphi = \beta - 3\beta sech^2(-\sqrt{\frac{-3\beta}{2}}\ln G)$, $G > 0$, $\beta = const < 0$.

To find a special solution into explicit form of the overdetermined system Eq. (3.3) we put $G = e^\psi V$, where the unknown linear function $\psi = \sum_{j=1}^n a_j x_j - \sigma t$, $\sigma \neq 0$, has real-valued coefficients and therefore $V(t,x)$ should satisfy

$$LV - 2((\vec{a}, \nabla_x V) + \sigma V_t) + (\sigma^2 + 1 - |a|^2)V+ \qquad (3.8)$$

$$(< V_t - \sigma V, (\nabla_x + \vec{a})V >, \vec{B}), \vec{a} = (a_1, \ldots, a_n),$$

$$\sum_{j=1}^n V_{x_j}^2 - V_t^2 + 2V((\vec{a}, \nabla_x V) + \sigma V_t) + V^2(|a|^2 - \sigma^2 - 1) = 0.$$

We shall assume further on that the following overdetermined system of 4 PDE holds:

$$\begin{vmatrix} LV = 0 \\ \sum_{j=1}^n V_{x_j}^2 - V_t^2 = 0 \quad \text{(eikonal equations)} \\ \sum_{j=1}^n a_j V_{x_j} + \sigma V_t = 0 \\ \sum_{j=1}^n b_j V_{x_j} + b_0 V_t = 0 \end{vmatrix} \qquad (3.9)$$

under the additional assumptions: $\sum_{j=1}^n a_j^2 = \sigma^2 + 1$, $\sum_{j=1}^n a_j b_j = b_0 \sigma$.

Evidently, Eq. (3.9) $\Rightarrow$ Eq. (3.8).

Consider now Eq. (3.5). We are looking for a solution having the form $G = \psi + W(t,x)$ and the linear function ψ is defined as above. Then Eq. (3.5) is rewritten as:

$$\begin{vmatrix} (\nabla_{t,x}\psi, \vec{B}) + (\nabla_{t,x}W, \vec{B}) + LW = 0 \\ \sum_{j=1}^n (a_j + W_{x_j})^2 - (W_t - \sigma)^2 = 1. \end{vmatrix} \qquad (3.10)$$

Suppose now that W satisfies

$$\begin{vmatrix} LW = 0 \\ \sum_{j=1}^n W_{x_j}^2 - W_t^2 = 0 \\ \sum_{j=1}^n a_j W_{x_j} + \sigma W_t = 0 \\ \sum_{j=1}^n b_j W_{x_j} + b_0 W_t = 0 \end{vmatrix} \qquad (3.11)$$

under the additional conditions $\sum_{j=1}^n a_j^2 = \sigma^2 + 1$, $\sum_{j=1}^n a_j b_j = \sigma b_0$. Certainly, Eq. (3.11) $\Rightarrow$ Eq. (3.10) and Eq. (3.9), Eq. (3.11) coincide.

Remark 3.1. Let $F \in C^2(\mathbf{R}^1)$ be arbitrary and $V = F(\alpha(t, x))$ verifies Eq. (3.9) for some $\alpha \in C^2$. Evidently, then $(F')^2 (\sum_{j=1}^n \alpha_{x_j}^2 - \alpha_t^2) = 0$, $F' L\alpha + F''(\alpha_t^2 - \sum_{j=1}^n \alpha_{x_j}^2) = 0$, $F'(\sum_{j=1}^n a_j \alpha_{x_j} + \sigma \alpha_t) = 0$ etc. This way we conclude that if α verifies the overdetermined system

$$\begin{vmatrix} L\alpha = 0 \\ \sum_{j=1}^n \alpha_{x_j}^2 - \alpha_t^2 = 0 \\ \sum_{j=1}^n a_j \alpha_{x_j} + \sigma \alpha_t = 0 \\ \sum_{j=1}^n b_j \alpha_{x_j} + b_0 \alpha_t = 0, \end{vmatrix} \qquad (3.12)$$

then for every $F \in C^2$ Eq. (3.9) with $V = F(\alpha)$ holds.

To solve Eq. (3.12) we look for $\alpha(t, x)$ of linear form, i.e. $\alpha(t, x) = -\sum_{j=1}^n x_j^0 x_j + t$, $x_j^0 = const.$

One gets immediately that

$$\sum_{j=1}^n x_j^0 = 1, \sum_{j=1}^n a_j x_j^0 = \sigma, \sum_{j=1}^n b_j x_j^0 = b_0 \qquad (3.13)$$

and moreover, $|a|^2 = \sum_{j=1}^n a_j^2 = \sigma^2 + 1$, $\sum_{j=1}^n a_j b_j = \sigma b_0$ $(\sigma \neq 0)$. If S_1^{n-1} is the unit sphere in $\mathbf{R}_x^n$ and B_1^n is the unit ball in $\mathbf{R}_x^n$ then the point $X^0 = (x_1^0, \ldots, x_n^0) \in S_1^{n-1}$, $a = (a_1, \ldots, a_n) \notin B_1^n$ and $X^0 \in \gamma_{1\sigma} \cap \gamma_2$, $\gamma_{1\sigma}$ and γ_2 being the hyperplanes $\sum_1^n a_j y_j = \sigma$, $\sum_1^n b_j y_j = b_0$ respectively. Put $b = (b_1, \ldots, b_n)$ and assume that $|b_0| < |b|$. Therefore, $|cos(\overrightarrow{a}, \overrightarrow{b})| < 1$, i.e. $\overrightarrow{a}, \overrightarrow{b}$ are not colinear and parts of $\gamma_{1\sigma}, \gamma_2$ are contained inside B_1^n.

Proposition 3.1. [109]

Consider Eq. (3.12) and suppose that: $|b_0| < |b|$, there exist a constant $\sigma \neq 0$, a vector $a \in \mathbf{R}^n$ such that: $|a|^2 = \sigma^2 + 1$, $(\overrightarrow{a}, \overrightarrow{b}) = b_0 \sigma$ and $\gamma_{1\sigma} \cap \gamma_2 \cap S_1^{n-1} \neq \{\emptyset\}$. Then Eq. (3.12) possesses infinitely many solutions depending on an arbitrary smooth function. It follows that Eq. (3.1) possesses infinitely many solutions written into explicit form: $u = \varphi(e^\psi F(\alpha(t, x)))$.

Remark 3.2. In many cases points $X^0 \in S_1^{n-1} \cap \gamma_{1\sigma} \cap \gamma_2$ do not exist for some $\sigma \neq 0$. Let $n \geq 3$, $\vec{b} \neq 0$ and $b_0 = 0 \Rightarrow \vec{a} \perp \vec{b}$, the point $P^0 = \frac{\sigma}{\sigma^2+1}a \in intB_1^n$ and $P^0 \in \gamma_{1\sigma} \cap \gamma_2$. The plane of codimension 2 $\gamma_{1\sigma} \cap \gamma_2$, $\sigma \neq 0$ will cross S_1^{n-1}, certainly. If $n = 3$ $\gamma_{1\sigma} \cap \gamma_2$ is a straight line crossing S_1^{n-1} at two points only. Otherwise, it is a smooth set of codimension 2 at S_1^{n-1}.

We shall not discuss Eq. (3.5), respectively then $u = \varphi(\psi + F(\alpha(t, x))$.

3.3   K-P equation and X, Y shallow water waves in the oceans

The K-P equation is given by the formula:

$$(u_t + 6uu_x + u_{xxx})_x + \alpha u_{yy} = 0, \tag{3.14}$$

$u = u(t, x, y)$, $\alpha^2 = 1$.

Later on we shall deal with $\alpha = 1$. By using Hirota's method [29], [63] Satsuma proved in [122] the existence of N-soliton solution of Eq. (3.14) having the form:

$$u = 2(log\ f)_{xx}, f = \sum_{\mu=0,1} exp\left[\sum_{1 \leq i < j}^{N} \mu_i \mu_j A_{ij} + \sum_{i=1}^{N} \mu_i \eta_i\right], \tag{3.15}$$

where $\eta_i = k_i(x + p_i y - C_i t)$, $C_i = k_i^2 + p_i^2$, $e^{A_{ij}} = \frac{3(k_i - k_j)^2 - (p_i - p_j)^2}{3(k_i + k_j)^2 - (p_i - p_j)^2}$. (See also Chapter 6.)

We do not have resonances if $e^{A_{ij}} \neq 0$. Resonances appear if for some (i, j): $e^{A_{ij}} = 0$. Thus, $N = 1 \Rightarrow u = \frac{k_1^2}{2} sech^2 \frac{\eta_1}{2}$, $sech\ x = \frac{2}{e^x + e^{-x}}$.

Let $N = 2$. Resonance exists iff $\sqrt{3}(k_1 - k_2) = \pm(p_1 - p_2)$. Further on we shall take $sign$ " $+$ " in front of $p_1 - p_2$, assuming $p_1 > p_2 > 0$ $\Rightarrow k_1 > k_2 > 0$.

The case of triple resonance $N = 3$

$$\sqrt{3}(k_1 - k_2) = \pm(p_1 - p_2)$$

$$\sqrt{3}(k_1 - k_3) = \pm(p_1 - p_2)$$

can be investigated in a similar way as the case of resonance for $N = 2$. As it is more complicated we discuss it in Chapter 6 (6.4).

Suppose now that $N = 2$ and $e^{A_{12}} \neq 0$ (no resonance). Then the corresponding solution of Eq. (3.14) given by formula Eq. (3.15) becomes

$$u =$$

$$2\frac{k_1^2 e^{\eta_1} + k_2^2 e^{\eta_2}}{(1 + e^{\eta_1} + e^{\eta_2} + e^{A_{12}+\eta_1+\eta_2})^2} \tag{3.16}$$

$$+ 2\frac{e^{\eta_1+\eta_2}[(k_1 - k_2)^2 + e^{A_{12}}(k_1 + k_2)^2 + k_2^2 a^{A_{12}+\eta_1} + k_1^2 e^{A_{12}+\eta_2}]}{(1 + e^{\eta_1} + e^{\eta_2} + e^{A_{12}+\eta_1+\eta_2})^2}$$

and it is called X wave ($u(0) > 0$). Fix η_1, y. Then $u \sim \frac{k_1^2}{2} sech^2 \frac{\eta_1 + A_{12}}{2}$ for $t \to \infty$, $u \sim \frac{k_1^2}{2} sech^2 \eta_1$ for $t \to -\infty$.

As it concerns the resonance case for $N = 2$, the solution is written as:

$$u = 2\frac{k_1^2 e^{\eta_1} + k_2^2 e^{\eta_2} + (k_1 - k_2)^2 e^{\eta_1+\eta_2}}{(1 + e^{\eta_1} + e^{\eta_2})^2}, u(0) > 0 \tag{3.17}$$

and is called Y wave.

Exercise 1. Consider the function

$$f(x, y) = \frac{x + k^2 y + (1 - k)^2 xy}{(1 + x + y)^2}, \quad x, y \geq 0, \quad 0 < k < 1$$

and k is parameter. Then $f(x, y) \leq \frac{1}{4}$ and $f(x, y) = \frac{1}{4} \iff x = 1, y = 0$; $lim \ sup_{(x,y) \to (\infty,\infty)} f(x, y)$ can be studied easily.

Hint. Consider two cases a) $y = 0$ and b) $y > 0$. In case a) $f(x, 0) < \frac{1}{4}$ for $x \neq 1$, $x \geq 0$ and $f(1, 0) = \frac{1}{4}$. In case b) fix $y > 0$, $x \geq 0$ and consider the quadratic polynomial in $k \in [0, 1]$ $f_{xy}(k) = \frac{x + k^2 y + (1-k)^2 xy}{(1+x+y)^2}$. The coefficient in front of k^2 is $\frac{y(x+1)}{(1+x+y)^2} > 0$, i.e. $f_{xy}(k)$ is strictly convex and therefore $f_{xy}(k) < max(f_{xy}(0), f_{xy}(1))$. As $f_{xy}(1) \leq 1/4$ according to a), one must prove only that $f_{xy}(0) \leq 1/4$. Show that $f_{xy}(0) \leq \frac{1}{4} \iff 0 \leq (1+y-x)^2$. For this nice proof we are indebted to N. Nikolov and A. Ivanov.

As usually, we shall study the profiles of the waves for $t = 0$, $t = \pm\frac{1}{4}, \pm\frac{1}{2}, \pm\frac{3}{4}, \pm 1$ etc. We shall concentrate at $t = 0$ as the other cases are treated in a similar way.

Thus, $N = 2$, no resonance case, and denote by $l_1 : \eta_1 = 0$, $l_2 : \eta_2 = 0$ the straight lines passing through the origin in $0xy$. Put for l an arbitrary line through 0. Then $u|_{l_1}$ and $u|_{l_2}$ are kinks-antikinks, while $u|_l$ is a soliton if $l \neq l_1, l_2$. Monotonically increasing (decreasing) bdd function $v(s)$ on $\mathbf{R}^1$ is called kink (antikink) — see [78], [110] if it possesses two horizontal asymptotes $v = \alpha$, $v = \beta$, $\alpha < (>)\beta$. In our considerations here we assume that kinks possess two horizontal asymptotes $v = \alpha$, $v = \beta$ at $\pm\infty$ but $v(s)$ is not obliged to be strictly monotone everywhere. Those are generalized kinks. In the case $l_1, l_2 : \alpha > 0$. The definition of soliton is standard.

One can easily see that

$$lim_{y \to -\infty} u|_{l_1} = \frac{2k_1^2 e^{A_{12}}}{(1 + e^{A_{12}})^2} = \alpha_1$$

$$lim_{y \to +\infty} u|_{l_1} = \frac{k_1^2}{2} = \beta_1$$

$$lim_{y \to +\infty} u|_{l_2} = 2\frac{k_2^2 e^{A_{12}}}{(1 + e^{A_{12}})^2} = \beta$$

$$lim_{y \to -\infty} u|_{l_2} = \frac{k_2^2}{2} = \alpha$$

X waves are formed by $u|_{l_1}$ and $u|_{l_2}$.

In the resonance case a new wave appears, namely $l_3 : x = -\alpha_0 y$, $\alpha_0 = p_1 + k_2\sqrt{3}$, while $l_1 : x + p_1 y = 0$, $l_2 : x + p_2 y = 0$. More precisely, the linear functions $\eta_1 = \eta_2 \iff x + \alpha_0 y = 0$, i.e. $\eta_1|_{l_3} = \eta_2|_{l_3}$. We put $l_1^+ = l_1 \cap \{y \geq 0\}$, $l_{2,3}^- = l_{2,3} \cap \{y \leq 0\}$. Then $u > 0$, $u|_{l_1}$, $u|_{l_2}$, $u|_{l_3}$ are kinks with a horizontal asymptote at $u = \alpha = 0$ and the second one at β_j, $j = 1, 2, 3$.

Moreover, $lim_{y \to \infty} u|_{l_1} = \frac{k_1^2}{2} = \beta_1$, $lim_{y \to -\infty} u|_{l_2} = \frac{k_2^2}{2} = \beta_2$, while $lim_{y \to -\infty} u|_{l_3} = \frac{(k_1 - k_2)^2}{2} = \beta_3$. In other words, the resonance gives rise of a new born wave kink with a maximal amplitude $\frac{(k_1 - k_2)^2}{2} = \beta_3$, $\beta_3 < \beta_1$ but $\beta_3 < \beta_2 \iff k_1 < 2k_2$. If $0 \in l \neq l_1, l_2, l_3$ is a straight line in $0xy$ then $u|_l$ is a soliton. l_1^+, l_2^- and l_3^- form the configuration Y wave (see Fig. 3.1).

Both X, Y waves can be observed in the oceans (even in the Mediterranean sea) during lowtides. We propose below a geometrical interpretation of the Y wave and pictures of X, Y waves taken from Mediterranean sea on May 25, 2014 (see Fig. 3.2).

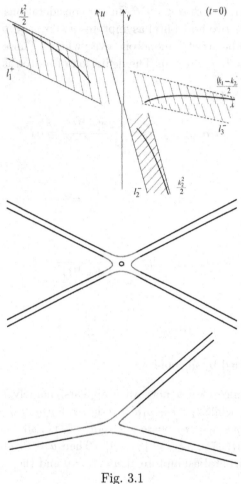

Fig. 3.1

3.4   Solutions of first order linear and cubic nonlinear first order hyperbolic pseudodifferential equations

This section is devoted to Eq. (3.18), Eq. (3.19), where:

$$(D_t - c|D_x|)u = f(t, x) \in D'(\mathbf{T}_t^1 \times \mathbf{T}_x^n) \qquad (3.18)$$

with a solution $u \in D'(\mathbf{T}_t^1 \times \mathbf{T}_x^n)$. As usual $\mathbf{T}_x^n$ stands for the n-dimensional 2π torus, $c \in \mathbf{R}^1$.

The cubic nonlinear first order hyperbolic Eq. (3.19) is given by the

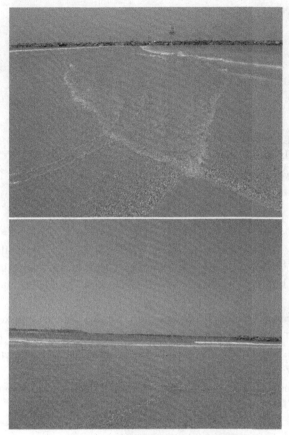

Fig. 3.2

formula:

$$(-D_t + |D_x|)u = u|u|^2 \tag{3.19}$$

with $x \in \mathbf{T}^n$, $t \in [0, T]$, $0 < T$ being possibly sufficiently small, $D_t = \frac{1}{i}\partial_t$. As we know each $L^2(\mathbf{T}^n_x)$ function $f(x)$ can be developed in Fourier series: $f \to \sum_{\alpha \in \mathbf{Z}^n} a_\alpha e^{i<\alpha, x>}$.

We define the Ψdo $|D_x|$ as follows:

$$|D_x|f = \sum_{\alpha \in \mathbf{Z}^n} a_\alpha |\alpha| e^{i<\alpha, x>},$$

the series being convergent in distribution sense in $D'(\mathbf{T}^n)$. We introduce now the function

$$f_{\geq 0} = P_{\geq 0}(f) = \sum_{\alpha \in \mathbf{Z}_+^n} a_\alpha e^{i<\alpha,x>}, \qquad (3.20)$$

$\mathbf{Z}_+ = N \cup \{0\}$. A function u satisfying the equation

$$(-D_t + |D_x|)u_{\geq 0} = P_{\geq 0}(|u_{\geq 0}|^2 u_{\geq 0}) \qquad (3.21)$$

is called Szegö solution of Eq. (3.19).

Having in mind that $f(x,t) = \sum_{(\tau,\alpha) \in \mathbf{Z}^{n+1}} a_{\tau,\alpha} e^{i(t\tau+<\alpha,x>)}$ if $f \in D'(\mathbf{T}^{n+1})$ we look for a solution of Eq. (3.18) of the form $u = \sum_{(\tau,\alpha) \in \mathbf{Z}^{n+1}} u_{\tau,\alpha} e^{i(t\tau+<\alpha,x>)}$, i.e.

$$(\tau - c|\alpha|)u_{\tau,\alpha} = f_{\tau,\alpha}, \forall (\tau,\alpha) \in \mathbf{Z}^{n+1}. \qquad (3.22)$$

Thus, $\tau_0 = c|\alpha_0|$ for some $(\tau_0,\alpha_0) \in \mathbf{Z}^{n+1} \Rightarrow f_{\tau_0,\alpha_0} = 0$, while $\tau \neq c|\alpha|$, $\forall (\tau,\alpha) \in \mathbf{Z}^{n+1} \Rightarrow u_{\tau,\alpha} = \frac{f_{\tau,\alpha}}{\tau - c|\alpha|}$. Equation (3.18) is nonlovable in $D'(\mathbf{T}^{n+1})$ if $\tau_0 = c|\alpha_0|$ but $f_{\tau_0,\alpha_0} \neq 0$; $c = \frac{\tau_0}{|\alpha_0|} \in Q \setminus 0 \iff |\alpha_0| \in \mathbf{N}, |\alpha_0| \notin \mathbf{N} \iff |\alpha_0| \notin Q$.

The operator $D_t - c|D|$ possesses an infinite dimensional kernel and is not $C^\infty(\mathbf{T}^{n+1})$ hypoelliptic if $\tau = c|\alpha|$ for infinitely many $(\tau,\alpha) \in \mathbf{Z}^{n+1}$. For example, $c = 1 \Rightarrow \tau = |\alpha| \to \tau^2 = |\alpha|^2$ and the Pythagorean numbers are infinitely many.

Assume now that $c^2 > 0$ satisfies the small denominators condition [14]:

$$|c^2 - \frac{p}{q}| \geq \frac{K}{|q|^{2+\sigma}} \qquad (3.23)$$

for each $p, q \in \mathbf{Z} \setminus 0$ and for some $\sigma > 0$, $K = K(c^2,\sigma) > 0$. Let $c > 0$. Then Eq. (3.23) implies that $|\tau - c|\alpha|| \geq \frac{\tilde{K}}{(|\alpha|+|\tau|)^{2\sigma+3}}$, $(\tau,\alpha) \in \mathbf{Z}^{n+1} \setminus 0$, $\tilde{K} = const > 0$.

Proposition 3.2.

For almost all $c \in \mathbf{R}^1$ in the sense of Lebesgue measure the operator $D_t - c|D_x|$ is C^∞, analytic and Gevrey hypoelliptic on $\mathbf{T}^{n+1}$.

The Cauchy problem for Eq. (3.18) with initial condition $u_0(x)$ can be easily studied in $D'([0,T) \times \mathbf{T}_x^n)$ as then $f = \sum_\alpha f_\alpha(t)e^{i<\alpha,x>}$, $u = \sum_\alpha u_\alpha(t)e^{i<\alpha,x>}$, $u_0(x) = \sum u_{0\alpha}e^{i<\alpha,x>}$ and therefore

$$\left| \begin{array}{l} u'_\alpha(t) - ic|\alpha|u_\alpha(t) = if_\alpha(t) \\ u_\alpha(0) = u_{0\alpha}. \end{array} \right. \qquad (3.24)$$

Our second step is to investigate Eq. (3.21) (see [50], [51]). We look for a solution having the form

$$u = \frac{a(t)}{1 - c(t)e^{i[<\alpha,x>+\beta t]}}, 0 \neq \alpha \in \mathbf{Z}^n, \beta \in \mathbf{R}^1. \qquad (3.25)$$

If $|c(t)| < 1$ the geometric progression formula shows that $u = u_{\geq 0} = P_{\geq 0}(u)$. Put $z = e^{i[<\alpha,x>+\beta t]}$, $w = c(t)z$. From Eq. (3.25) we obtain that

$$i\frac{\partial u}{\partial t} = \frac{ia'}{1-w} + ia\frac{(c'+i\beta c)z}{(1-w)^2}, \qquad (3.26)$$

where $a(t)$, $c(t)$, α and β are unknown, $|c| < 1$.

On the other hand,

$$|D_x|\left(\frac{1}{1-cz}\right) = \sum_{k=1}^{\infty} c^k(t)|D_x|(z^k) = \sum_{k=1}^{\infty} c^k(t)|\alpha|kz^k.$$

As $|w| = |c| < 1$ we have

$$\sum_{k=0}^{\infty} w^k = \frac{1}{1-w}, \sum_{k=1}^{\infty} kw^{k-1} = \frac{1}{(1-w)^2} \Rightarrow \sum_{k=1}^{\infty} kw^k = \frac{w}{(1-w)^2}.$$

This way we conclude that

$$|D_x|\left(\frac{a(t)}{1-w}\right) = |\alpha|\frac{aw}{(1-w)^2}. \qquad (3.27)$$

Combining Eq. (3.26), Eq. (3.27) we get for $\beta = |\alpha|$:

$$i\frac{\partial u}{\partial t} + |D_x|u = \frac{ia'(t)}{1-w} + \frac{ia(t)c'(t)z}{(1-w)^2} = \qquad (3.28)$$

$$i\left(\frac{a}{c}\right)'\frac{c}{1-cz} + i\frac{a}{c}\frac{c'}{(1-cz)^2}.$$

V. Georgiev, N. Tzvetkov and N. Visciglia have shown the following algebraic lemma in [50].

Lemma 3.1.

Consider the function $\frac{1}{1-cz}$, $|c| < 1$, $c \in \mathbf{C}^1$, $z = e^{i\Theta}$, $\Theta \in [0, 2\pi]$. *Then*

$$P_{\geq 0}\left(\left(\frac{1}{1-cz}\right)^2\frac{1}{1-\bar{c}\bar{z}}\right) = \frac{1-|c|^2cz}{(1-cz)^2(1-|c|^2)^2} = \frac{a|a|^2}{(1-|c|^2)(1-cz)^2} +$$

$$\frac{a|a|^2|c|^2}{(1-|c|^2)^2(1-cz)} = \frac{h(z) - h(\bar{c})}{z - \bar{c}}, h = \frac{z}{(1-cz)^2}.$$

Due to Eq. (3.28) and the Lemma equation Eq. (3.21) leads to the ODE system

$$\left|\begin{array}{l} i\left(\frac{a}{c}\right)' = \frac{a}{c}\frac{|a|^2|c|^2}{(1-|c|^2)^2} \\ ic' = \frac{c|a|^2}{1-|c|^2}. \end{array}\right. \tag{3.29}$$

Thus, with some $c_0 \in (0,1)$ the function $c(t) = c_0 e^{-ip_1 t}$, $p_1 = \frac{a_0^2}{1-c_0^2} > 0$, $c(0) = c_0$, $a(0) = a_0 \neq 0$ and $a(t) = a_0 e^{-ip_2 t}$, $p_2 = \frac{a_0^2}{(1-c_0^2)^2}$ satisfy Eq. (3.29).

Proposition 3.3.

For each $\alpha \in \mathbf{Z}_+^n$ and $\beta = |\alpha|$ the equation Eq. (3.21) possesses the solution

$$u_{\geq 0} = \frac{a_0 e^{-ip_2 t}}{1 - c_0 e^{i[<\alpha,x>+t(|\alpha|-p_1)]}}.$$

3.5   Possible generalizations of Proposition 3.3

Consider the same Eq. (3.21) and look for a solution having the form

$$u_{\geq 0} = \sum_{m=1}^{N} \frac{a_m(t)}{1 - c_m(t)z}, z = e^{i[<\alpha,x>+\beta t]},$$

$$0 \neq \alpha \in \mathbf{Z}_+^n, \beta = |\alpha|, 0 < |c_1(0)| < |c_2(0)| < \ldots < |c_N(0)| < 1.$$

Then

$$\left(i\frac{\partial}{\partial t} + |D_x|\right)u_{\geq 0} = i\sum_{m=1}^{N}\left(\frac{a_m}{c_m}\right)'\frac{c_m}{1-c_m z} + i\sum_{m=1}^{N}\frac{a_m}{c_m}\frac{c_m'}{(1-c_m z)^2},$$

$$|c_m(t)| < 1, 1 \leq m \leq N.$$

Evidently,

$$|u_{\geq 0}|^2 u_{\geq 0} = u_{\geq 0}^2 |\bar{u}_{\geq 0}| = \sum_{j,k=1}^{N}\frac{a_j^2 \bar{a}_k}{(1-c_j z)^2(1-\bar{c}_k \bar{z})} +$$

$$2\sum_{1\leq j<k\leq N}\sum_{l=1}^{N}\frac{a_j a_k \bar{a}_l}{(1-c_j z)(1-c_k z)(1-\bar{c}_l \bar{z})}.$$

Moreover, $c_j(0) \neq c_k(0)$ for $j < k$; for $l \neq j, k$ $c_k(0) \neq \bar{c}_l(0)$, $c_j(0) \neq \bar{c}_l(0)$. Certainly, we must find $P_{\geq 0}(|u_{\geq 0}|^2 u_{\geq 0})$. We observe that

$P_{\geq 0}\left(\frac{1}{(1-c_j z)^2}\frac{1}{1-\bar{c}_k \bar{z}}\right) = P_{\geq 0}\left(\frac{z}{(1-c_j z)^2}\frac{1}{z-\bar{c}_k}\right) = \frac{f_j(z)-f_j(\bar{c}_k)}{z-\bar{c}_k}$, where $f_j(z) = \frac{z}{(1-c_j z)^2}$.

In fact, $\frac{1}{z-\bar{c}_k} = \sum_{p=0}^{\infty}\frac{\bar{c}_k^p}{z^{p+1}}$, on the other hand,

$$P_{\geq 0}\left(\frac{1}{1-c_j z}\frac{1}{1-c_k z}\frac{1}{1-\bar{c}_l \bar{z}}\right) =$$

$$P_{\geq 0}\left(\frac{z}{(1-c_j z)(1-c_k z)}\frac{1}{z-\bar{c}_l}\right) = \frac{g_{jk}(z)-g_{jk}(\bar{c}_l)}{z-\bar{c}_l},$$

where $g_{jk}(z) = \frac{z}{(1-c_j z)(1-c_k z)}$, $c_j \neq c_k$. Certainly, $\frac{z}{(1-c_j z)(1-c_k z)} -$ $\frac{\bar{c}_l}{(1-c_j \bar{c}_l)(1-c_k \bar{c}_l)} = \frac{z-\bar{c}_l}{c_j-c_k}\left[\frac{c_j}{(1-c_j z)(1-c_j \bar{c}_l)} - \frac{c_k}{(1-c_k z)(1-c_k \bar{c}_l)}\right]$. As in the previous case, we compare the coefficients participating in the left-hand side and in the right-hand side of Eq. (3.21) and in front of $\frac{1}{1-c_m z}$, $\frac{1}{(1-c_m z)^2}$, $1 \leq m \leq N$.

The corresponding complex system of ODE takes the form:

$$\left|\begin{array}{l} i\left(\frac{a_m}{c_m}\right)' c_m = P_m(a,c,\bar{a},\bar{c}), 1 \leq m \leq N \\ ic_m \frac{a_m}{c_m} = Q_m(a,c,\bar{a},\bar{c}), \quad 1 \leq m \leq N, \end{array}\right.$$

Q_m, P_m being algebraic functions of the arguments $a = (a_1,\ldots,a_N)$, $c = (c_1,\ldots,c_N)$, $\bar{a} = (\bar{a}_1,\ldots,\bar{a}_N)$.

Separating the real and imaginary parts of a_j, c_j, P_m, Q_m we obtain a real-valued system of 4N ODE in normal form with 4N unknown functions $Re\, a_j$, $Im\, a_j$, $Re\, c_j$, $Im\, c_j$. Taking the Cauchy data $c_j(0)$, $a_j(0)$ such that $0 < |c_1(0)| < \ldots < |c_N(0)| < 1$, $a_j(0) \neq 0$. $1 \leq j \leq N$ we construct a local in t solution, i.e. $|t| \leq T$, $0 < T \ll 1$. Unfortunately, it does not have the elegant form proposed in Proposition 3.3. Put $\tilde{P}_N = \prod_{m=1}^{N}(z - \frac{1}{c_m(t)})$, $P_j(z,t) = \frac{\tilde{P}_N(z,t)}{z-\frac{1}{c_j}(t)}$, $\tilde{Q}(z,t) = (-1)^N \sum_{m=1}^{N}\frac{a_m(t)}{c_m(t)}P_j(z,t)$. Then the local in t solution $u_{\geq 0} = \frac{\tilde{Q}(z,t)}{\tilde{P}_N(z,t)}$, where $\tilde{Q}$, $\tilde{P}_N$ are polynomials in z of degrees $N-1$, N respectively having coefficients depending on $a_m(t)$, $c_m(t)$ or on $\frac{1}{c_m(t)}$, $1 \leq m \leq N$ only. Therefore, we can find rational solution $u_{\geq 0}$ of Eq. (3.21) and $z = e^{i[<\alpha,x>+|\alpha|t]}$, $0 \neq \alpha \in \mathbf{Z}_+^n$.

3.6   Exact solutions of Tzitzeica equation

Consider the equation

$$(ln\, v)_{xy} = v - \frac{1}{v^2}, \qquad (3.30)$$

i.e. $v = e^u$ satisfies

$$u_{xy} = e^u - e^{-2u} \tag{3.31}$$

($\Longleftrightarrow u_{\alpha\alpha} - u_{\beta\beta} = 4(e^u - e^{-2u})$, where $x = \alpha + \beta$, $y = \alpha - \beta$).

Equation (3.30) arises from differential geometry and for the first time was considered by the rumanian mathematician G. Tzitzeika in 1910 (see [128], [129]). There are different ways in finding solutions of special form of Eq. (3.30) (respectively Eq. (3.31)). Below we propose three of them.

(A) Put $u = ln|\varphi(x) + \psi(x)|$, where $\varphi(x)$, $\psi(x)$ are unknown functions in the equation

$$u_{xx} - u_{yy} = 4(e^u - e^{-2u}).$$

Then $u_{xx} - u_{yy} = 4(\varphi + \psi - \frac{1}{(\varphi+\psi)^2}) = \frac{(\varphi+\psi)(\varphi''-\psi'')-((\varphi')^2-(\psi')^2)}{(\varphi+\psi)^2}$, i.e.

$$(\varphi + \psi)(\varphi'' - \psi'') - ((\varphi')^2 - (\psi')^2) = 4(\varphi + \psi)^3 - 4.$$

We are looking for

$$(\varphi')^2 = a_0\varphi^3 + a_1\varphi^2 + a_2\varphi + a_3 \equiv P_3(\varphi),$$

$$(\psi')^2 = b_0\varphi^3 + b_1\varphi^2 + b_2\varphi + b_3 \equiv Q_3(\psi),$$

the coefficients a_i, b_j being unknown.

Certainly, $\varphi'' = \frac{1}{2}P_3'(\varphi)$, $\psi'' = \frac{1}{2}Q_3'(\psi)$ and therefore, $\frac{\varphi+\psi}{2}(P_3'(\varphi) - Q_3'(\psi)) - P_3 + Q_3) = 4(\varphi + \psi)^3 - 4$. Comparing the coefficients in front of $\varphi^k.\psi^l$, $0 \le k + l \le 3$ and having in mind that a_i, b_j participate in a linear way in the corresponding expressions we get that $a_0 = -b_0 = 8$, $a_1 = b_1 = A$, $a_2 = -b_2 = C_1$, $a_3 = C_2$, $b_3 = C_2 - 4$, A, C_1, C_2 being arbitrary constants. So

$$\left| \begin{array}{l} (\varphi')^2(x) = 8\varphi^3 + A\varphi^2 + C_1\varphi + C_2 \\ (\psi')^2(y) = -8\psi^3 + A\psi^2 - C_1\psi + C_2 - y. \end{array} \right.$$

(B) Suppose that Eq. (3.31) possesses a solution having the form of traveling wave: $u = \varphi(x + cy)$, $c = const \ne 0$; $\xi = x + cy$. Thus,

$$c\varphi''(\xi) = e^\varphi - e^{-2\varphi} \Rightarrow \frac{c}{2}(\varphi')^2 = e^\varphi + \frac{1}{2}e^{-2\varphi} + C,$$

$0 \le C = const$. The standard change $z = e^\varphi > 0$ leads to

$$(z')^2 = \frac{2}{c}z^3 + \frac{2C}{c}z^2 + \frac{1}{c} = P_3(z) = \frac{2}{c}\tilde{P}_3(z). \tag{3.32}$$

To fix the ideas, let $c > 0$ and $C = -\frac{3}{2}$. Then $P_3(z) = \frac{2}{c}(z^3 - \frac{3}{2}z^2 + \frac{1}{2})$ $= \frac{2}{c}(z - 1)^2(z + \frac{1}{2})$, $-\frac{1}{2} < z < 1$, i.e. we can take $z' = \sqrt{P_3(z)} = \sqrt{\frac{2}{c}(1 - }$

$z)\sqrt{\frac{1}{2}+z}$. Having in mind that $(arc\ sech\sqrt{x})' = \frac{-1}{2x\sqrt{1-x}}$ for $0 < x < 1$, where $sech\lambda = \frac{1}{ch\lambda} = \frac{2}{e^{\lambda}+e^{-\lambda}}$ we get that $z(\xi) = 1 - \frac{3}{2}sech^2(\frac{1}{2}\sqrt{\frac{3}{c}}\xi + C_0)$ [44]. Evidently, $z(\pm\infty) = 1$. Moreover, $-\frac{1}{2} < z < 1$, while $z > 0 \iff \sqrt{\frac{2}{3}} > sech(\frac{1}{2}\sqrt{\frac{3}{c}}\xi + C_0)$, i.e. for some $0 < \eta_0 = arc\ sech\sqrt{\frac{2}{3}}$ we have: $z(\xi)$, $\eta = \frac{1}{2}\sqrt{\frac{3}{2}}\xi + C_0$ is positive for $|\eta| > \eta_0$. In other words, soliton solution of Eq. (3.32) gives a solution of Tzitzeica equation $\varphi = \ln z(\xi)$ outside some compact interval on the real line. At the end points of that interval ξ_1, ξ_2 $z(\xi)$ vanishes and therefore $\lim_{\xi\to\xi_1}\varphi(\xi) = -\infty$, $\lim_{\xi\to\xi_2}\varphi(\xi) = -\infty$, $\varphi' > 0$ for $\xi > \xi_2$, $\varphi'(\xi) < 0$ for $\xi < \xi_1$, $\varphi(\pm\infty) = 0$.

Suppose now that $c < 0$. According to Viete's formulas for $\tilde{P}_3 = z^3 + Cz^2 + \frac{1}{2}$ we have that $z_1 + z_2 + z_3 = -C$, $z_1z_2 + z_1z_3 + z_2z_3 = 0$, $z_1z_2z_3 = -\frac{1}{2}$. We are interested when (for which values of C) $z_1 < 0 < z_2 < z_3$. Thus, let $z_1 < 0$. Consequently, z_2 and z_3 are roots of the equation $w^2 - \frac{1}{2z_1^2}w - \frac{1}{2z_1} = 0$, i.e. $z_{2,3} = \frac{\frac{1}{2z_1^2} \pm \sqrt{\frac{1}{4z_1^4}+\frac{2}{z_1}}}{2}$. We must assume that $\frac{1}{4z_1^4} + \frac{2}{z_1} > 0 \iff 1 + (2z_1)^3 > 0 \iff 1 > -2z_1$, i.e. for $-\frac{1}{2} < z_1 < 0$ the roots $0 < z_2 < z_3$. Moreover, then $-C = z_1 + \frac{1}{2z_1^2} = \frac{2z_1^3+1}{2z_1^2} > 0$, as $\frac{3}{4} < 1 + 2z_1^3 < 1$. Evidently, $z_1 \to 0 \Rightarrow C \to -\infty$. As it is well known the equation $(z')^2 = \frac{2}{c}\tilde{P}_3(z) = P_3(z)$ possesses for $c < 0$ and $z_2 < z < z_3$ a periodic solution $z(\xi)$ with the additional conditions $z(\alpha_1) = z_2$, $z'(\alpha_1) = 0$, $z(\alpha_1 + \frac{T}{2}) = z_3$, $z'(\alpha_1 + \frac{T}{2}) = 0$, where the period $\frac{T}{2} = \int_{z_2}^{z_3}\frac{d\lambda}{\sqrt{P_3(\lambda)}} > 0$.

The corresponding solution of Tzitzeica equation $\varphi = \ln z(\xi)$ is C^{∞} smooth and periodic with the same period T. Put z_1, z_2, z_3 for the zeros of $\tilde{P}_3(z)$. If z_1, z_2, z_3 are real then $z_1z_2z_3 = -1$ and either one of them is negative (say $z_1 < 0$) and the other two are positive ($0 < z_2 < z_3$) or $z_1 \leq z_2 \leq z_3 < 0$. If $z_2 = \bar{z}_3$ and z_1 is simple root we have that $z_1 < 0$. One can see that a bounded positive solution $z(\xi)$ exists if and only if it is located between two positive roots of $P_3(z)$ (say $0 < z_2 < z_3$) such that $P_3(z) > 0$ for $z_2 < z < z_3$. Otherwise either unbounded solution exists or soliton type solution ($z_1 < 0$, $z_2 = z_3 > 0$, $P_3(z) > 0$ for $z \in (z_1, z_2)$) of Eq. (3.32) exists which does not generate a global solution on $\mathbf{R}^1$ of Eq. (3.31).

Proposition 3.4.

The only global bounded solutions (traveling waves) of Eq. (3.31) are the periodic ones. The periodic solution of Eq. (3.32) can be expressed by the Jacobi elliptic function cn, namely,

$$z(\xi) = z_2 + (z_3 - z_2)cn^2 \left[\xi \sqrt{\frac{z_3 - z_2}{2|c|}}, s \right],$$

where $c < 0$, $s^2 = \frac{z_3 - z_2}{z_3 - z_1} \in (0,1)$ *and the period of* z $T = 2\sqrt{\frac{2|c|}{z_3 - z_2}} \int_0^1 \frac{dq}{[(1-s^2q^2)(1-q^2)]^{1/2}}$. *Certainly,* $z_1 < 0 < z_2 < z_3$ *and* $\varphi(\xi) = ln\, z(\xi)$.

(C)   We are looking now for multisolution type solutions of Eq. (3.30) imitating the Hirota approach (see for more details [63], [75]).

The first observation is that if the functions $f(x,y)$, $g(x,y)$ satisfy the system

$$\left| \begin{array}{l} f^2 (ln\, f)_{xy} - f^2 + g^2 = 0 \\ g^2 (ln\, g)_{xy} + fg - g^2 = 0, \end{array} \right. \tag{3.33}$$

then the fraction $v = \frac{f}{g}$ is a solution of Eq. (3.30).

In fact, Eq. (3.33) implies that

$$\left| \begin{array}{l} (ln\, f)_{xy} - 1 + \frac{g^2}{f^2} = 0 \\ (ln\, g)_{xy} + \frac{f}{g} - 1 = 0, \end{array} \right. \tag{3.34}$$

i.e.

$$\left(ln\frac{f}{g} \right)_{xy} + \frac{g^2}{f^2} - \frac{f}{g} = 0 \Rightarrow (ln\, v)_{xy} + \frac{1}{v^2} - v = 0.$$

The change $g = \tau^2$ in Eq. (3.34) (second equation) leads to

$$\tau^2 2(ln\tau)_{xy} + f - \tau^2 = 0 \Rightarrow f = \tau^2(1 - 2(ln\tau)_{xy})$$

and from the first equation of Eq. (3.33) we get

$$\tau^4 (1 - 2(ln\tau)_{xy})^2 ((ln\, f)_{xy} - 1) + \tau^4 = 0,$$

i.e.

$$(1 - 2(ln\tau)_{xy})^2 [(ln(\tau^2 - 2(\tau_{xy}\tau - \tau_x\tau_y)))_{xy} - 1] + 1 = 0. \tag{3.35}$$

This way we obtain an analogue of the Hirota bilinear form (see Chapter 6)

$$(1 - 2(ln\tau)_{xy})^2 [(ln(\tau^2 - 2\tau_{xy}\tau + 2\tau_x\tau_y))_{xy} - 1] + 1 = 0. \tag{3.36}$$

Conclusion: A class of solutions of Eq. (3.30) can be found in the form $v = \frac{f}{g} = 1 - 2(ln\tau)_{xy} = e^u \iff u = ln(1 - 2(ln\tau)_{xy})$. To find a N soliton solution of Eq. (3.36) we are looking for

$$\tau_N = 1 + \sum_{1 \le i \le N} f_i + \sum_{1 \le i_1 \le i_2 \le N} c_{i_1 i_2} f_{i_1} f_{i_2} +$$

$$\sum_{1\le i_1 < i_2 < i_3 \le N} c_{i_1 i_2 i_3} f_{i_1} f_{i_2} f_{i_3} + \ldots + c_{1\ldots N} f_1 f_2 \ldots f_N,$$

where $f_i = e^{k_i x + m_i y + s_i}$, k_i, s_i are arbitrary constants, m_i depends on k_i only and the constants $c_{i_1 i_2}, \ldots, c_{12\ldots N}$ are expressed by $k_1, \ldots, k_N$.

Assume that $N = 1$, i.e. $\tau_1 = 1 + e^{kx + my + s}$. Putting τ_1 in Eq. (3.36) we obtain that for $m = \frac{3}{k}$ the function τ_1 satisfies Eq. (3.36). With some technical efforts one can see that $\tau_2 = 1 + f_1 + f_2 + p_{12} f_1 f_2$ satisfies Eq. (3.36) for $f_i = e^{k_i x + 3y/k_i + s_i}$, where k_i, s_i are arbitrary constants and

$$p_{12} = \frac{(k_1 - k_2)^2 (k_1^2 - k_1 k_2 + k_2^2)}{(k_1 + k_2)^2 (k_1^2 + k_1 k_2 + k_2^2)}. \tag{3.37}$$

To find out p_{12} we remark that after putting τ_2 in Eq. (3.36) we obtain a polynomial with respect to f_1 and f_2 in the left-hand side. In fact, τ, τ_x, τ_y, $\tau_{xy}, \ldots$ are polynomials of f_1, f_2 of second degree, $(ln\tau)_{xy} = \frac{\tau \tau_{xy} - \tau_x \tau_y}{\tau^2}$ and in the numerator and denominator stand polynomials of 4 degree with respect to f_1, f_2. As it concerns

$$(ln(\tau^2 - 2\tau\tau_{xy} + 2\tau_x\tau_y))_{xy} = \left(\frac{2\tau\tau_x - 2\tau_x\tau_{xy} - 2\tau\tau_{xxy} + 2\tau_{xx}\tau_y + 2\tau_x\tau_y}{(\tau^2 - 2\tau\tau_{xy} + 2\tau_x\tau_y)} \right)_y,$$

i.e. in the numerator and in the denominator participate polynomials of f_1, f_2 of degree 8. Multiplying Eq. (3.36) by $\tau^4(\tau^2 - 2\tau\tau_{xy} + 2\tau_x\tau_y)^2$ we get a polynomial in the left-hand side of the variables f_1, f_2. Equalizing to zero the coefficients of that polynomial of f_1, f_2 we obtain an algebraic (nonlinear) and complicated system that should be satisfied by k_1, k_2, $p_{12}(k_1, k_2)$. There is among these algebraic equations a linear one that is satisfied by Eq. (3.37). The overdetermined system is satisfied for that choice of p_{12}, as one can see after hard computations.

In the paper [75] it is shown that

$$\tau_N = 1 + \sum_{i=1}^{N} f_i + \sum_{m=2}^{N} \left(\sum_{1 \le i_1 < i_2 < \ldots < i_m \le N} f_{i_1} \ldots f_{i_m} \prod_{1 \le j < r \le m} p(k_{i_j}, k_{i_r}) \right). \tag{3.38}$$

As the proof is very technical, we omit it here (see [75] for details).

Example 2.

1). $N = 3 \Rightarrow$

$$\tau_3 = 1 + f_1 + f_2 + f_3 + p_{12} f_1 f_2 + p_{13} f_1 f_3 + p_{23} f_2 f_3 + p_{12} p_{13} p_{23} f_1 f_2 f_3,$$

$f_i = e^{k_i x + 3y/k_i + s_i}$, $1 \le i \le 3$, $p_{ij} = p(k_i, k_j) = \frac{(k_i - k_j)^2 (k_i^2 - k_i k_j + k_j^2)}{(k_i + k_j)^2 (k_i^2 + k_i k_j + k_j^2)}$, $i \ne j$.

2). Assume that $k_1 = k_2 = k \ne 0$, i.e. $p_{12} = 0$, $f_1 = f_2 e^{s_1 - s_2}$ when a 2-soliton solution of the form Eq. (3.38) does not exist.

One can see that $\tau = 1 + c(x - \frac{3y}{k})^2 e^{kx+3y/k} - \frac{c^2}{12k^2} e^{2(kx+\frac{3y}{k})}$, $c \in \mathbf{R}^1$ satisfies Eq. (3.36), k being real.

3). Let $k_1 = ib$, $k_2 = -ib$, $b \neq 0$. Then $\tau = sin(bx - \frac{3y}{b}) + \sqrt{3}(bx + \frac{3y}{b})$ satisfies Eq. (3.36) for b real.

With the last 3 examples we complete that Chapter.

Chapter 4

First integrals of systems of ODE having jump discontinuities

4.1 Introduction

This chapter deals with the interaction of peakon- (anti) peakon and peakon-kink solutions of several equations of Mathematical Physics. More precisely, at first we consider the so-called b-evolution equation:

$$m_t + m_x u + bmu_x = 0, m = u - u_{xx},$$

i.e.

$$u_t - u_{xxt} + (b+1)uu_x = bu_x u_{xx} + uu_{xxx}$$

and its special cases: $b = 2$ — Camassa-Holm equation [31] and $b = 3$ — Degasperis-Procesi equation [41]. Then following [116] we investigate a generalization of the b-equation, namely an evolution PDE containing both quadratic and cubic nonlinearities:

$$m_t = bu_x + \frac{k_1}{2}[m(u^2 - u_x^2)]_x + \frac{k_2}{2}(2mu_x + m_x u), \qquad (4.1)$$

$m = u - u_{xx}$, b, k_1, k_2 — constants. It is also called a generalized Camassa-Holm equation. We shall use this terminology in Section 4.

Evidently, for $k_1 = 0$, $k_2 = -2$, $b = 0$ we obtain the Camassa-Holm equation and for $k_2 = 0$, $k_1 = -2$ — the so-called cubic nonlinear evolution equation ([46], [48], [103], [115]). In studying the collision of the waves (anti)peakons and kinks we look for solutions of the above mentioned equations in a special form (Ansatz). As it is well known now, this way the interaction of these waves is reduced to the solvability of a system of ODE having non-smooth right-hand side (usually, jumps appear). We are able in many cases to obtain global first integrals of that system reducing its solvability to the solvability of scalar nonlinear ODE satisfied by the components of the vector-solution. Under some restrictions these ODEs can be

solved in quadratures and in some special cases in elementary functions or
in elliptic ones. The bonification of that approach is double sided. In fact,
we can find solutions into explicit form and then study their interactions
via a classical method from the Analysis. In Chapter 5 of our book [110] we
considered in details the interaction of 2 peakons, satisfying the Camassa-
Holm equation. Relying on the paper [116] we shall find first integrals of
other systems of ODE enabling us to study the propagation, interaction
and collision of the equations of Mathematical Physics mentioned above.
The Lax representation, bi-Hamiltonian structure and infinitely many con-
servation laws for the generalized Camassa-Holm equation are avoided in
our investigations.

4.2   Interaction of 2 peakon solutions of the Camassa-Holm equation

As that book is devoted to the nonlinear waves, their interaction and propagation, we shall begin with the Camassa-Holm equation, describing some
classes of peakons and antipeakons. This is the Camassa-Holm equation

$$u_t - u_{xxt} + 3uu_x = 2u_x u_{xx} + uu_{xxx}. \tag{4.2}$$

If $u(x,t)$ is a solution of Eq. (4.2), then $-u(x,-t)$ is a solution too.
$u = ce^{-|x-ct|}$ is a peakon. Eq. (4.2) possesses a distribution solution having
the form (Ansatz):

$$u(x,t) = p_1(t)e^{-|x-q_1(t)|} + p_2(t)e^{-|x-q_2(t)|}, \tag{4.3}$$

p_1, p_2, q_1, q_2 being classical C^1 functions. By using the Schwartz distribution theory in [110] we proved that the functions $p_{1,2}$, $q_{1,2}$ satisfy the
system of ODE

$$
\begin{aligned}
p_1' &= p_1 p_2 sgn(q_1 - q_2)e^{-|q_1-q_2|} \\
p_2' &= p_1 p_2 sgn(q_2 - q_1)e^{-|q_1-q_2|} \\
q_1' &= p_1 + p_2 e^{-|q_1-q_2|} \\
q_2' &= p_2 + p_1 e^{-|q_1-q_2|}
\end{aligned}
\tag{4.4}
$$

The existence and interaction of two peakons (i.e. $p_1 > 0$, $p_2 > 0$) was
studied in [36], [110]. Here we shall investigate the collision of the pair
peakon-antipeakon (see also [19]). To simplify the things we assume that
$p_1 = -p_2$, $q_1 = -q_2$. Then Eq. (4.4) reduces to

$$p_1' = -2p_1^2 sgn q_1 e^{-2|q_1|}$$
$$q_1' = p_1(1 - e^{-2|q_1|}). \tag{4.5}$$

As it is shown in [31] the system Eq. (4.5) has the solutions $p_1 = cotgh\ t$, $q_1 = ln\ cosh\ t \geq 0$. As we know

$$p_1 \sim \begin{cases} 1, & t \to \infty \\ \frac{1}{t}, & t \to 0 \\ -1, & t \to -\infty \end{cases}, q_1 \sim \begin{cases} \frac{t^2}{2}, & t \to 0 \\ |t| - ln\ 2, & t \to \pm\infty, \end{cases}$$

q_1 — even, $-p_1(-t) = p_1(t)$, $q_1(t) = q_1(-t)$.

Therefore $z = q_1 > 0$ for $t \neq 0$ has two asymptotics: $z = |t| - ln\ 2$. Put I: $u = p_1(t)e^{-|x-q_1(t)|}$, II: $u = -p_2(t)e^{-|x+q_1(t)|}$. I is peakon for $t > 0$ and antipeakon for $t < 0$, while II is antipeakon for $t > 0$ and peakon for $t < 0$ (see Fig. 4.1).

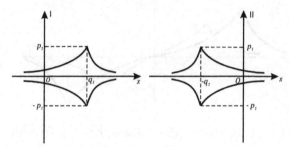

Fig. 4.1

Evidently, $|lim\ u|_{t\to 0, x\to 0} = \infty$, i.e. the collision between I and II should be at the moment $t = 0$, $(q_1(t), p_1(t))$ is a peak and $(-q_1(t), -p_1(t))$ is a trough for $t > 0$.

To study the interaction of I and II we shall consider 3 cases for $t > 0$ and fixed:

1). $x \leq -q_1(t) < 0 \Rightarrow x \leq -q_1(t) < q_1(t)$
2). $-q_1(t) < x \leq q_1(t)$
3). $x > q_1(t) > 0 \Rightarrow x > q_1 > -q_1$.

In case 1) $u = -2p_1(t)e^x sh\ q_1 < 0$, $u_x < 0$ and $r_1(t) = u|_{x=-q_1} = p_1(t)(e^{-2q_1} - 1) = -tgh\ t$, i.e. $0 < t_1 < t_2 \Rightarrow 0 > r_1(t_1) > r_1(t_2)$.

In case 2): $u = 2p_1(t)e^{-q_1} sh\ x \Rightarrow u_x > 0$, $u(x = 0, t) = 0$, $|u| \leq |tgh\ t| \leq 1$, $u(0,0) = 0$, $lim_{x,t\to 0} u_x = \infty$. $r_2 = u|_{x=q_1} = tgh\ t \to 1$, $t \to \infty$, $tgh\ t \sim t$ for $t \to 0$; $q_1 \sim \frac{t^2}{2}$ for $t \to 0$ and in case 3) $u = 2p_1(t)e^{-x} sh\ q_1$.

This way we come to Fig. 4.2.

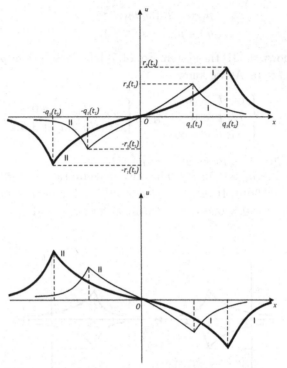

Fig. 4.2. i). (Before the collision, $t > 0$, $p_1(t) > 0$).
ii). (After the collision, $t < 0$, $p_1(t) < 0$).

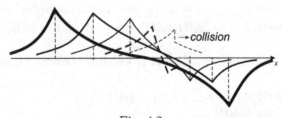

Fig. 4.2a.

To orientate in the situation near the origin assume that $-q_1 \leq x \leq q_1$, $q_1(t) \sim \frac{t^2}{2}$ near $t = 0$ and $x \sim 0 \Rightarrow sh\, x \sim x$. Therefore, $u \sim \frac{2}{t} e^{-\frac{t^2}{2}} x \sim \frac{2x}{t}$ and $|x| \leq \frac{t^2}{2}$. Geometrically we have Fig. 4.3 near 0:

The line segments tend to the origin where the solution is bounded;

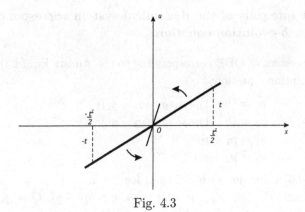

Fig. 4.3

$u(x = 0, t \neq 0) = 0$. Starting for some $t > 0$ and $x = -\infty$ with an antipeakon type wave II it transforms into peakon wave I due to their interaction. After the collision for $t = 0$ where u vanishes the waves overturn and the direction of the propagation too. We start at $x = +\infty$, $t < 0$ with the antipeakon I and go to $x = -\infty$ with the peakon II.

Possible generalizations

Hans Lundmark proposed the following formula for (anti)peakon solutions of Eq. (4.2) (see [79], [93], [94]):

$$p_1(t) = \frac{\lambda_1^2 b_1 + \lambda_2^2 b_2}{\lambda_1 \lambda_2 (\lambda_1 b_1 + \lambda_2 b_2)}$$

$$p_2(t) = \frac{b_1 + b_2}{\lambda_1 b_1 + \lambda_2 b_2}$$

$$q_1 = log \frac{(\lambda_1 - \lambda_2)^2 b_1 b_2}{\lambda_1^2 b_1 + \lambda_2^2 b_2}$$

$$q_3 = log(b_1 + b_2),$$

where $\lambda_1 \neq \lambda_2$, $\lambda_1 \lambda_2 \neq 0$ are real constants and $b_k(t) = c_k e^{\frac{t}{\lambda_k}}$, $c_k > 0$, $k = 1, 2$. In this case the geometrical picture is more exotic in comparison with Fig. 4.2 as then upright positions of the waves (solution Eq. (4.3) see Fig. 4.2a) appear. The calculations are more complicated.

Exercise 1. Solve the system Eq. (4.5) for $q_1 > 0$.

Hint. $p_1(e^{2q_1} - 1) - e^{2q_1} = C_1$ is a first integral of Eq. (4.5), $\frac{e^{2q_1}}{e^{2q_1} - 1} = \frac{1}{2}(1 + cotgh\, q_1)$, $\frac{1}{e^{2q_1} - 1} = \frac{1}{2}(cotgh\, q_1 - 1)$ etc.

4.3 First integrals of the dynamical system corresponding to the b-evolution equation

This is the system of ODE corresponding to the Ansatz Eq. (4.3) solutions of the b-evolution equation:

$$
\begin{aligned}
p_1' &= (b-1)p_1p_2 sgn\,(q_1-q_2)e^{-|q_1-q_2|} \\
p_2' &= (b-1)p_1p_2 sgn\,(q_2-q_1)e^{-|q_1-q_2|} \\
q_1' &= p_1 + p_2 e^{-|q_1-q_2|} \\
q_2' &= p_2 + p_1 e^{-|q_1-q_2|}.
\end{aligned}
\tag{4.6}
$$

Equation (4.6) coincides with Eq. (4.4) for $b = 2$.

The change $P = p_1 + p_2$, $p = p_1 - p_2$, $q = q_1 - q_2$, $Q = q_1 + q_2$, i.e. $p_1 p_2 = \frac{P^2 - p^2}{4}$ transforms Eq. (4.6) into

$$
\begin{aligned}
P' &= 0 \\
p' &= (b-1)\frac{P^2 - p^2}{2} sgn\,q\,e^{-|q|} \\
q' &= p(1 - e^{-|q|}) \\
Q' &= P(1 + e^{-|q|}).
\end{aligned}
\tag{4.7}
$$

At first we shall consider the case when $q \le 0 \Rightarrow -|q| = q$. The case $q \ge 0$ is studied in a similar way and we omit it. Thus, $P = const$ and assume that $b > 1$. Therefore,

$$
\frac{dp}{dq} = \frac{b-1}{2}\frac{(p^2 - P^2)e^q}{p(1 - e^q)} \Rightarrow p^2 - P^2 = \frac{A^{\frac{1}{b-1}}}{(1 - e^q)^{b-1}},
$$

$A = const > 0$ if $p^2 > P^2$. We obtained the first integral $(p^2 - P^2)(1 - e^q)^{b-1} = A$ of Eq. (4.7). Consequently,

$$
p' = \frac{b-1}{2}(p^2 - P^2)\left(1 - \frac{A^{\frac{1}{b-1}}}{(p^2 - P^2)^{\frac{1}{b-1}}}\right),
$$

i.e.

$$
p' = \frac{1}{2}(b-1)\frac{(p^2 - P^2)^{\frac{1}{b-1}} - A^{\frac{1}{b-1}}}{(p^2 - P^2)^{\frac{2-b}{b-1}}}.
\tag{4.8}
$$

In the case of Camassa-Holm equation $b = 2$ and Eq. (4.8) takes the form

$$
p' = \frac{1}{2}(p^2 - P^2 - A).
\tag{4.9}
$$

So $ln(B^{-1}|\frac{\sqrt{P^2+A}+p}{\sqrt{P^2+A}-p}|^{\frac{1}{\sqrt{P^2+A}}}) = -t \Rightarrow$

$$
\left|\frac{\sqrt{P^2 + A} + p}{\sqrt{P^2 + A} - p}\right| = B^{\sqrt{P^2+A}}e^{-t\sqrt{P^2+A}}, \quad B = const > 0.
\tag{4.10}
$$

Conclusion: $p(t)$ is a homographic function of the argument $e^{t\sqrt{P^2+A}}$:

$$p(t) = \frac{s_1 e^{t\sqrt{P^2+A}} + s_2}{s_3 e^{t\sqrt{P^2+A}} + s_4}, \; s_1, s_2, s_3, s_4 \text{ being appropriate real constants.}$$

Remark 4.1. $ln\frac{a+bx}{a-bx} = ln\frac{\frac{a}{b}+x}{\frac{a}{b}-x} = 2arctgh\frac{bx}{a}$ for $a^2 > b^2x^2$.

From the system Eq. (4.7) we have:

$$q' = p(1 - e^q) = p\frac{A}{p^2 - P^2} \Rightarrow q = A\int \frac{p(t)dt}{p^2(t) - P^2}.$$

From the first integral of the system found above we get: $e^q = 1 - \frac{A}{p^2 - P^2}$.
Therefore, $Q = Pt + P\int e^{q(t)}dt = 2Pt - AP\int \frac{dt}{p^2(t)-P^2}$.

Exercise 2. Find $\int \frac{p(t)dt}{p^2(t)-P^2}$ and $\int \frac{dt}{p^2(t)-P^2}$ via elementary functions (logarithms).

Hint. In the corresponding integrals with p homographic function of $e^{t\sqrt{P^2+A}}$ make the change $\lambda = e^{t\sqrt{P^2+A}}$. This way you will obtain integrals of rational functions having 2nd order polynomials in the numerators and the cubic polynomial $\lambda(A_1(\lambda) - PB_1(\lambda)) \times (A_1(\lambda) + PB_1(\lambda))$ in the denominator, where $A_1(\lambda) = \lambda s_1 + s_2$, $B_1(\lambda) = \lambda s_3 + s_4$, $P = const$. For appropriate constants $s_1 \neq 0$, s_2, $s_3 \neq 0$, s_4 the polynomial possesses 3 simple roots $\lambda_1 = 0$, λ_2, λ_3 etc.

In the case $b = 3$ the equation Eq. (4.8) takes the form

$$I = \int \frac{dp}{p^2 - P^2 - \sqrt{A}\sqrt{p^2 - P^2}} = t, p^2 > P^2.$$

The Euler substitution $p^2 - P^2 = \lambda^2(p - P)^2$, i.e. $p = P\frac{\lambda^2+1}{\lambda^2-1}$ leads to
$I = 2\int \frac{d\lambda}{\sqrt{A}\lambda^2 - 2P\lambda - \sqrt{A}}$

$$= 2\begin{cases} \frac{1}{\sqrt{-\Delta}}arctg\frac{\sqrt{A}\lambda - P}{\sqrt{-\Delta}}, & \Delta < 0, \Delta = P^2 - A \\ \frac{1}{2\sqrt{\Delta}}ln|\frac{\sqrt{A}\lambda - P - \sqrt{\Delta}}{\sqrt{A}\lambda - P + \sqrt{\Delta}}|, & \Delta > 0 \end{cases}$$

(see [44]).
Therefore,

$$\left|\frac{\sqrt{A}\lambda - P - \sqrt{\Delta}}{\sqrt{A}\lambda - P + \sqrt{\Delta}}\right| = Ce^{\sqrt{\Delta}t} \text{ for } \Delta > 0. \tag{4.11}$$

Again we have that λ is a homographic function of $\mu = e^{\sqrt{\Delta}t}$, i.e. $\lambda = \frac{s_5\mu + s_6}{s_7\mu + s_8} = \frac{A_2(\mu)}{B_2(\mu)}$. One can see that $\lambda = \sqrt{\frac{p+P}{p-P}}$ then implies

$$p(t) = P\frac{A_2^2 + B_2^2}{A_2^2 - B_2^2}, A_2 = A_2(e^{\sqrt{\Delta}t}), B_2 = B_2(e^{\sqrt{\Delta}t}). \tag{4.12}$$

From Eq. (4.7) we have that $e^q = 1 - \frac{\sqrt{A}}{\sqrt{p^2 - P^2}}$, where $p^2 - P^2 = P^2 \frac{4A_2^2 B_2^2}{(A_2^2 - B_2^2)^2}$ etc. As it concerns Q, $Q' = P(1 + e^q) \Rightarrow$

$$Q = 2Pt - \sqrt{A} \int \frac{dt}{\sqrt{p^2 - P^2}}, \quad \sqrt{p^2 - P^2} = 2P \frac{A_2 B_2}{|A_2^2 - B_2^2|}. \qquad (4.13)$$

The integral $\int \frac{dt}{\sqrt{p^2 - P^2}}$ can be computed via the change $\mu = e^{\sqrt{\Delta} t}$ that reduces it to an integral of a rational function (fraction of two polynomials of order 2). Thus, the solutions of Degasperis-Procesi equation via the Ansatz Eq. (4.3) are found.

Remark 4.2. It is interesting to know when the solutions of the system Eq. (4.6) are Hamiltonian ones. It is easy to see that Eq. (4.6) is Hamiltonian system iff $b = 2$ (i.e. Camassa-Holm equation). The simple proof is given below.

The system Eq. (4.7) is Hamiltonian one if there exists a function $H(q, Q, p, P)$ such that

$$0 = P' = -\frac{\partial H}{\partial Q} \Rightarrow H = H(q, p, P)$$

$$\frac{\partial H}{\partial P} = Q' = P(1 + e^{-|q|})$$

$$-\frac{\partial H}{\partial q} = p' = (b-1)\frac{P^2 - p^2}{2} sgn\, q e^{-|q|}$$

$$\frac{\partial H}{\partial p} = q' = p(1 - e^{-|q|}) \Rightarrow H = \frac{p^2}{2}(1 - e^{-|q|}) + \varphi(q, P).$$

Thus, $\frac{\partial H}{\partial P} = P(1 + e^{-|q|}) = \frac{\partial \varphi}{\partial P} \Rightarrow \varphi = \frac{P^2}{2}(1 + e^{-|q|}) + g(q)$.
Therefore, $H(q, p, P) = \frac{p^2 + P^2}{2} + e^{-|q|}\frac{P^2 - p^2}{2} + g(q)$.
We conclude that $sgn q e^{-|q|}\frac{P^2 - p^2}{2} - g'(q) = -\frac{\partial H}{\partial q} = (b-1)\frac{P^2 - p^2}{2} sgn q e^{-|q|}$, i.e. $(b-2) sgn q e^{-|q|}\frac{P^2 - p^2}{2} = -g'(q) \Rightarrow g(q) = (b-2)e^{-|q|}\frac{P^2 - p^2}{2}$ which is impossible for $b \neq 2$ as q, p are independent variables. The Hamiltonian of Camassa-Holm equation is $H(q, p, P) = \frac{p^2 + P^2}{2} + e^{-|q|}\frac{P^2 - p^2}{2}$.

4.4 First integrals of the ODE system corresponding to the Ansatz Eq. (4.3) solutions of the generalized Camassa-Holm Eq. (4.1)

The solutions we are looking for satisfy Eq. (4.1) in generalized sense with $b = 0$. In Chapter 5 of our book [110] it was proved that the generalized (distribution) solution of the classical Camassa-Holm Eq. (4.2) satisfies the dynamical system Eq. (4.4). In a similar way we can see that if u has the form Eq. (4.3) then p_1, p_2, q_1, q_2 are solutions of the following system of ODE:

$$
\begin{aligned}
p_1' &= -\tfrac{1}{2}k_2 p_1 p_2 \, sgn(q_1 - q_2) e^{-|q_1 - q_2|} \\
p_2' &= -\tfrac{1}{2}k_2 p_1 p_2 \, sgn(q_2 - q_1) e^{-|q_1 - q_2|} \\
q_1' &= -k_1 p_1 p_2 e^{-|q_1 - q_2|} - \tfrac{1}{3}k_1 p_1^2 - \tfrac{1}{2}k_2(p_1 + p_2 e^{-|q_1 - q_2|}) \\
q_2' &= -k_1 p_1 p_2 e^{-|q_1 - q_2|} - \tfrac{1}{3}k_1 p_2^2 - \tfrac{1}{2}k_2(p_2 + p_1 e^{-|q_1 - q_2|}).
\end{aligned}
\tag{4.14}
$$

The same change as in the system Eq. (4.6) transforms Eq. (4.14) into

$$
\begin{aligned}
P' &= 0 \Rightarrow P = const. \\
p' &= -k_2 \frac{P^2 - p^2}{4} sgn \, q e^{-|q|} \\
Q' &= -k_1 \frac{P^2 - p^2}{2} e^{-|q|} - \tfrac{1}{6}k_1(P^2 + p^2) - \tfrac{1}{2}k_2 P(1 + e^{-|q|}).
\end{aligned}
\tag{4.15}
$$

To solve the system we need at least one first integral of Eq. (4.15). Assume that $q \le 0 \Rightarrow e^{-|q|} = e^q$, $sgn \, q = -1$, $k_1, k_2 > 0$.
Then

$$
\frac{dp}{dq} = \frac{k_2}{4} \frac{(-p^2 + P^2)e^q}{p(\tfrac{1}{3}k_1 P + \tfrac{k_2}{2}(1 - e^q))}, P = const.
$$

and consequently

$$
(P^2 - p^2)\left(\frac{1}{3}k_1 P + \frac{k_2}{2}(1 - e^q)\right) = A = const.
$$

is a first integral of Eq. (4.15) and $e^q = 1 - (\frac{A}{P^2 - p^2} - \tfrac{1}{3}k_1 P)\frac{2}{k_2}$. This way we obtain for $p(t)$ the ODE:

$$
p' = \frac{k_2}{4}(P^2 - p^2)\left(1 - \frac{2}{k_2}\frac{A}{P^2 - p^2} + \frac{2}{3}\frac{k_1}{k_2}P\right).
\tag{4.16}
$$

Thus

$$
II = \int \frac{dp}{(P^2 - p^2)c_1 + c_2} = t + const., c_1 = \frac{k_2}{4}(1 + \frac{2}{3}\frac{k_1}{k_2}P), c_2 = -\frac{A}{2}, A \ne 0.
$$

Remark 4.3. In the paper [116] explicit formulas to the solutions of the dynamical system Eq. (4.15) are given. For sake of completeness we include them here.

Suppose that $0 < 1 + \frac{2}{3}\frac{k_1}{k_2}P = \Gamma \leq 1$. Then Eq. (4.15) admits the following solution: $P(t) = P = const$,

$$p(t) = \pm a_2 \frac{1 + A_3 e^{Bt}}{1 - A_3 e^{Bt}}, q(t) = \pm ln \frac{4\Gamma a_2^2 A_3 e^{Bt}}{a_2^2(1 + A_3 e^{Bt})^2 - P^2(1 - A_3 e^{Bt})^2}$$

$$Q(t) = ln \left| \frac{A_3 e^{Bt} - \frac{P - a_2}{P + a_2}}{A_3 e^{Bt} - \frac{P + a_2}{P - a_2}} \right| - \frac{2k_1 a_2^2(3\Gamma - 1)}{3B(A_3 e^{Bt} - 1)} -$$

$$\frac{1}{2}[k_2 P + \frac{1}{3}k_1(P^2 + a_2^2)]t + D,$$

where $a_2 > |P|$, $A_3 > 0$, $B = -\frac{1}{2}a_2 k_2 \Gamma$, D is an arbitrary constant.

As the formula for the solution in the case $\Gamma > 1$ is similar we omit it.

Exercise 3. Show that $u = Ce^{-|x - ct|}$ is a weak solution of the generalized Camassa-Holm equation with $b = 0$ if $k_1 \neq 0$, the constant c (velocity of the wave) is such that $\Delta = 3k_2^2 - 16k_1 c \geq 0$ and the amplitude $C = \frac{-\sqrt{3}(\sqrt{3}k_2 \pm \sqrt{\Delta})}{4k_1}$.

Exercise 4. [116]. Consider the cubic nonlinear equation

$$m_t = \frac{k_1}{2}[m(u^2 - u_x^2)]_x, m = u - u_{xx},$$

i.e. in the generalized Camassa-Holm equation Eq. (4.1) we have put $b = 0$, $k_2 = 0$. The corresponding dynamical system Eq. (4.14) then is satisfied by the solution Eq. (4.3). Equation (4.14) takes the form

$$\begin{aligned} &p_1' = 0, p_2' = 0 \Rightarrow p_1 = const, p_2 = const \\ &q_1' = -k_1 p_1 p_2 e^{-|q_1 - q_2|} - \frac{1}{3}k_1 p_1^2 \\ &q_2' = -k_1 p_1 p_2 e^{-|q_1 - q_2|} - \frac{1}{3}k_1 p_2^2. \end{aligned} \quad (4.17)$$

Solve the system Eq. (4.17). Can you propose geometrical interpretation of two peakon interaction and collision for $t = 0$ if $p_1^2 < p_2^2$?

Hint. Consider the cases: a). $p_1^2 = p_2^2$, b). $p_1^2 \neq p_2^2$; The ODE $q'(t) = me^{-a|t|} + n$, $m, n = const$, $a > 0$, $q(0) = 0$, has the solution $q(t) = -sgn(t)\frac{m}{a}(e^{-a|t|} - 1) + nt$.

Answer: b). $q_1(t) = sgnt\frac{3k_1 p_1 p_2}{|k_1(p_1^2 - p_2^2)|}(e^{-\frac{1}{3}|k_1(p_1^2 - p_2^2)t|} - 1) - \frac{1}{3}k_1 p_1^2 t$, $q_2(t) = sgnt\frac{3k_1 p_1 p_2}{|k_1(p_1^2 - p_2^2)|}(e^{-|\frac{1}{3}k_1(p_1^2 - p_2^2)t|} - 1) - \frac{1}{3}k_1 p_2^2 t$.

Below we propose an extract from the theory of the Cross-coupled Camassa-Holm equation [37] where not only the positions of the peakons under consideration are taken into account but their canonical moments as well. The peakon dynamics is governed by a Hamilton function, i.e. the solutions of some Hamilton equation should be found. We formulate as an exercise the simplest possible case of such a system.

Exercise 5. Consider the Hamilton system

$$
\begin{aligned}
q' &= \tfrac{\partial H}{\partial m} = -ne^r sh\, q \\
m' &= mne^r \cosh q = -\tfrac{\partial H}{\partial q} \\
r' &= \tfrac{\partial H}{\partial n} = -me^r sh\, q \\
n' &= -\tfrac{\partial H}{\partial r} = mne^r sh\, q
\end{aligned}
\tag{4.18}
$$

with Hamiltonian $H = -nme^r sh\, q$. $H = H_0$ is a first integral, of course. Put $q(0) = q_0 > 0$, $r(0) = r_0$, $m(0) = m_0$, $n(0) = n_0 \neq 0$. Find the solutions of the Cauchy problem to Eq. (4.18).

Hint. $n' = -H_0 \Rightarrow n = -H_0 t + n_0$; $\frac{r'}{n'} = \frac{dr}{dn} = -\frac{1}{n} \Rightarrow r = ln \frac{n_0 e^{r_0}}{n_0 - H_0 t}$. From the first equation it follows that $q' = -n_0 e^{r_0} sh\, q$ and therefore $q = ln \frac{e^{q_0}+1+(e^{q_0}-1)e^{-n_0 r_0 t}}{e^{q_0}+1-(e^{q_0}-1)e^{-n_0 r_0 t}}$ ($\int \frac{dq}{sh\, q} = ln \left| \frac{e^q - 1}{e^q + 1} \right|$). From $m = -\frac{H_0}{ne^r sh\, q}$ it follows that

$$
m = -\frac{H_0[(e^{q_0} + 1)^2 e^{n_0 r_0 t} - (e^{q_0} - 1)^2 e^{-n_0 r_0 t}]}{2n_0 e^{r_0}(e^{2q_0} - 1)}.
$$

What about the life-span of the solutions of Eq. (4.18)? What about their blow up?

Exercise 6. (a) Show that the function

$$
u = C sgn(x - ct)(e^{-|x-ct|} - 1), C \neq 0
\tag{4.19}
$$

is a smooth kink.

(b) Find the necessary and sufficient conditions under which Eq. (4.19) satisfies the generalized Camassa-Holm Eq. (4.1).

Hint. Consider the cases $x - ct > 0$ and $x - ct < 0$ separately.

Answer. $k_2 = 0$, $c = -\frac{1}{2}b$, $k_1 C^2 + b = 0$. If we are working in the real domain $bk_1 < 0$. As b is a fixed constant in the generalized Camassa-Holm equation the speed of the weak kink Eq. (4.18) is fixed: $c = -\frac{1}{2}b$. Certainly, $b \neq 0$.

Remark 4.4. The Ansatz Eq. (4.3) deals with two peakon solutions of the generalized Camassa-Holm equation Eq. (4.1) with $b = 0$. It can be generalized to the case of N-peakon solutions, i.e. we are looking for u having the following form:

$$
u(x,t) = \sum_{j=1}^{N} p_j(t) e^{-|x-q_j(t)|}.
$$

In the paper [116] it is shown that the amplitudes $p_j(t)$ and the positions $q_j(t)$ must satisfy rather complicated system of ODE with discontinuous right-hand side (jump type singularities appear there). For the sake of completeness and having in mind that the formula could be useful for further investigations we proposed it here. Thus,

$$
\begin{aligned}
p_j' &= -\tfrac{1}{2}k_2 p_j \sum_{k=1}^{N} p_k sgn(q_j - q_k)e^{-|q_j - q_k|} \\
q_j &= -\tfrac{1}{2}k_2 \sum_{k=1}^{N} p_k e^{-|q_j - q_k|} + \tfrac{1}{2}k_1(\tfrac{1}{3}p_j^2 + \\
&\quad \sum_{i,k=1}^{N} p_i p_k (1 - sgn(q_j - q_i) \times sgn(q_j - q_k))e^{-|q_j - q_i| - |q_j - q_k|}.
\end{aligned}
\tag{4.20}
$$

The difficulties in the solvability of Eq. (4.20) increase with N. They are outside the scope of this book.

Exercise 7. Solve the ODE $q'(t) = f(t) + A(t)e^{q(t)}$.

Hint. Make at first the change $z = e^q$ obtaining this way a Bernoulli first order ODE.

Answer. $q = \int f(t)dt - ln|C - \int A(t)e^{\int f(t)dt}dt|$, $C = const$.

4.5 Interaction of kink-peakon solutions to the generalized Camassa-Holm equation. First integral

In [116] the following Ansatz of solution to the generalized Camassa-Holm Eq. (4.1) is proposed:

$$
u = p_1(t)sgn(x - q_1(t))(e^{-|x - q_1(t)|} - 1) + p_2(t)e^{-|x - q_2(t)|}. \tag{4.21}
$$

It is proved there that if $k_2 = 0$ and b is arbitrary then the amplitudes $p_{1,2}(t)$ and the position functions $q_{1,2}$ satisfy the following system of ODE:

$$
\begin{aligned}
p_1 &= \pm\sqrt{-\tfrac{b}{k_1}} \quad (\text{eventually, } kb_1 < 0) \\
p_2' &= k_1 p_1^2 p_2 sgn(q_2 - q_1)e^{-|q_1 - q_2|} \\
q_1' &= -\tfrac{1}{2}b - k_1 p_1 p_2 sgn(q_2 - q_1)e^{-|q_1 - q_2|} \\
q_2' &= -\tfrac{1}{3}k_1 p_2^2 - \tfrac{1}{2}k_1 p_1^2 + k_1(p_1^2 - p_1 p_2 sgn(q_2 - q_1))e^{-|q_1 - q_2|} + \\
&\quad k_1 sgn(q_2 - q_1)p_1 p_2.
\end{aligned}
\tag{4.22}
$$

In the interaction of single kink and N-peakons the Ansatz for the solution u is similar to Eq. (4.21) (again $k_2 = 0$, $b \neq 0$):

$$
u = p_0(t)sgn(x - q_0(t))(e^{-|x - q_0(t)|} - 1) + \sum_{j=1}^{N} p_j e^{-|x - q_j|}. \tag{4.23}
$$

In Eq. (4.23) $p_0 = \pm\sqrt{-\tfrac{b}{k_1}}$, $q_0' = \tfrac{1}{2}k_1 p_0^2 + k_0 p_0 \sum_{i=1}^{N} p_i sgn(q_0 - q_i)e^{-|q_0 - q_i|}$. The ODE satisfied by p_j, q_j, $1 \leq j \leq N$ are very complicated and we omit their formulation (see [116]).

To fix the ideas, assume that in Eq. (4.22) $q_2 > q_1$. Put $B_1 = \pm\sqrt{-\frac{b}{k_1}}$, $B_2 = k_1 p_1 = \pm sgn\, k_1\sqrt{-bk_1}$. Then Eq. (4.22) takes the form

$$\begin{aligned}
p_2' &= -bp_2 e^{q_1-q_2} \\
q_1' &= -\tfrac{1}{2}b - B_2 p_2 e^{q_1-q_2} \\
q_2' &= -\tfrac{1}{3}k_1 p_2^2 + \tfrac{b}{2} - (b + B_2 p_2)e^{q_1-q_2} + B_2 p_2.
\end{aligned} \tag{4.24}$$

From the first two equations we have that

$$B_2 p_2 - bq_1 = \frac{bt}{2} + A_1, A_1 = const, \tag{4.25}$$

while the second and the third equations imply that

$$q' = -b + \frac{1}{3}k_1 p_2^2 - B_2 p_2 + be^q, \tag{4.26}$$

where $q = q_1 - q_2 < 0$.

We can apply the result of Exercise 7 to Eq. (4.26). Instead of, we make the change $q = log\, f(p_2)$. Therefore, $q'(t) = \frac{f'(p_2)p_2'}{f(p_2)} = -f'(p_2)bp_2$, the function f being unknown, $0 < f(p_2) < 1 \iff q < 0$. The Eq. (4.26) is transformed in

$$-bf'(p_2)p_2 = -b + \frac{1}{3}k_1 p_2^2 - B_2 p_2 + bf(p_2). \tag{4.27}$$

Equation (4.27) is a linear ODE with respect to the unknown function f and the independent variable p_2:

$$f'(p_2) = \frac{1}{p_2} - \frac{k_1}{3b}p_2 + \frac{B_2}{b} - \frac{f(p_2)}{p_2}. \tag{4.28}$$

Consequently,

$$f(p_2) = \frac{A_2}{p_2} + 1 - \frac{k_1}{9b}p_2^2 + \frac{B_2}{2b}p_2, A_2 = const \tag{4.29}$$

and $(e^{q_1-q_2} + \frac{k_1}{9b}p_2^2 - \frac{B_2}{2b}p_2 - 1)p_2 = A_2$ is a first integral of Eq. (4.24). We assume that $-\frac{k_1}{9b}p_2^2 + \frac{B_2}{2b}p_2 + 1 + \frac{A_2}{p_2} \in (0,1)$. In order to find $p_2(t)$ we will solve the ODE $p_2' = -bp_2 e^q = -bp_2 f(p_2)$, i.e.

$$I = \int \frac{dp_2}{-bA_2 - bp_2 + \frac{k_1}{9}p_2^3 - \frac{B_2}{2}p_2^2} = \int dt; t + C = I(p_2).$$

Now everything depends on the behavior of the zeroes of the cubic polynomial $P_3(p_2) = \frac{k_1}{9}p_2^3 - \frac{B_2}{2}p_2^2 - bp_2 - bA_2$. For example, if $P_3(p_2) = 0$ has 3 simple zeroes $\lambda_1, \lambda_2, \lambda_3$ then $I = c_1|p_2 - \lambda_1| + c_2|p_2 - \lambda_2| + c_3|p_2 - \lambda_3|$, c_i being real. If $P_3(p_2) = 0$ has one simple zero and 2 complex-valued then $I = c_1|p_2 - \lambda_1| +$ logarithmic expression of second order non vanishing polynomial of $p_2 + c_2$ multiplied by *arctg* of a linear function of p_2. The inversion of the function $p_2 = I^{-1}(t + C)$ is rather delicate — non explicit and it is better to work with definite integrals: $\int_{p_2^0}^{p_2} \frac{d\lambda}{P_3(\lambda)} = t - t_0$ etc.

Practically the system Eq. (4.24) is solved as $q_1 = \frac{1}{b}(B_2 p_2 - \frac{b^2}{2}t - A_1)$ and $q_2 = q_1 - log\, f(p_2)$.

Below we give a qualitative picture of the interaction and the collision between kink and peakon waves described by the generalized Camassa-Holm equation (see Fig. 4.4).

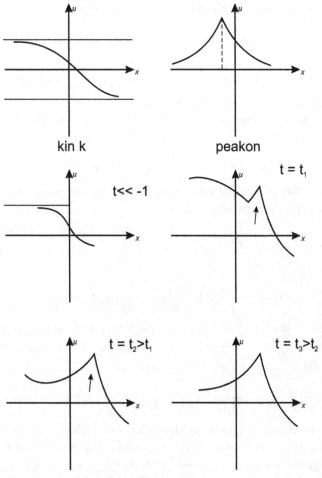

Fig. 4.4

4.6 Concluding remarks

It is interesting to find out soliton solutions of the generalized Camassa-Holm Eq. (4.1). To do this we look for a solution having the form $u = \Phi(x - ct)$, $\xi = x - ct$. Substituting the expression for u in the equation after one integration with respect to ξ one obtains the following second order ODE:

$$\Phi'' \left(c + \frac{1}{2}k_2\Phi + \frac{1}{2}k_1\Phi^2 - \frac{1}{2}k_1(\Phi')^2 \right) = \tag{4.30}$$

$$\left((b+c)\Phi + \frac{3}{4}k_2\Phi^2 + \frac{1}{2}k_1\Phi^3 \right) - \left(\frac{k_2}{4} + \frac{1}{2}k_1\Phi \right)(\Phi')^2.$$

Put $P_2(\Phi) = \frac{c + \frac{1}{2}k_2\Phi + \frac{1}{2}\Phi^2}{-\frac{1}{2}k_1}$, $P_3(\Phi) = \frac{(b+c)\Phi + \frac{3}{4}k_2\Phi^2 + \frac{1}{2}k_1\Phi^3}{-\frac{1}{2}k_1}$, $P_1(\Phi) = \frac{\frac{k_2}{4} + \frac{k_1}{2}\Phi}{\frac{k_1}{2}}$,
$k_1 \neq 0$. Then after the change $p = \Phi'(\xi)$, $\Phi'' = p\frac{dP}{d\Phi} = \frac{1}{2}\frac{d}{d\Phi}q$, $p^2 = q$, $k_1 \neq 0$
Eq. (4.30) can be rewritten as:

$$\frac{dq}{d\Phi} = 2\frac{P_3(\Phi) + qP_1(\Phi)}{q + P_2(\Phi)}. \tag{4.31}$$

The equation Eq. (4.31) is known as Abel ODE of second kind (see [74]). We are looking for a first integral of Eq. (4.31): $H(\Phi, q) = Aq^2 + B(\Phi)q + C(\Phi)$, where $A = const$ and B, C are smooth functions. As by definition of first integral $H(\Phi, q(\Phi)) \equiv const$, we obtain

$$q' = -\frac{B'q + C'}{2Aq + B} = \frac{-\frac{\partial H}{\partial \Phi}}{\frac{\partial H}{\partial q}}. \tag{4.32}$$

Equalizing the coefficients in front of q^2, q, q^0 in Eq. (4.31), Eq. (4.32) we get the overdetermined system $2A = 1$, $B = P_2$, $2P_1 = -B'$, $2P_3 = -C'$.
Fortunately, $B' = -2P_1 \equiv P_2'$. Therefore

$$H(\Phi, q) = \frac{1}{2}q^2 + \frac{c + \frac{1}{2}k_2\Phi + \frac{1}{2}k_1\Phi^2}{-\frac{1}{2}k_1}q + 2\frac{(b+c)\frac{\Phi^2}{2} + \frac{1}{4}k_2\Phi^3 + \frac{1}{8}k_1\Phi^4}{\frac{1}{2}k_1},$$

i.e.

$$-\frac{1}{4}k_1q^2 + \left(c + \frac{1}{2}k_2\Phi + \frac{1}{2}k_1\Phi^2\right)q - 2\left((b+c)\frac{\Phi^2}{2} + \frac{1}{4}k_2\Phi^3 + \frac{k_1\Phi^4}{8}\right) \equiv h = const \tag{4.33}$$

is the first integral.

There are different methods for solving the Abel equation but it is not possible in the general case to express $q(\Phi)$ by quadrature (see [74]). On the other hand, in our case $p^2 = q \Rightarrow p = \pm\sqrt{q(\Phi)} = \Phi'(\xi)$, i.e. $\int \frac{d\Phi}{\pm\sqrt{q(\Phi)}} = \xi + const \Rightarrow \xi = \pm\xi(\Phi)$.

From Eq. (4.33) we have that $q^2 + 2P_2(\Phi)q + P_4(\Phi) \equiv h$, where $P_4 = -4 \int P_3(\Phi)d\Phi$. Therefore, $q_{1,2}(\Phi) = P_2(\Phi) \pm \sqrt{P_2^2 - P_4(\Phi) + h}$.

Several restrictions on the coefficients k_1, k_2, b, c will enable us to solve $\int \frac{d\Phi}{\sqrt{q(\Phi)}}$ in elementary functions. In fact, $\sqrt{q_1(\Phi)} = \sqrt{P_2 + \sqrt{P_2^2 - P_4(\Phi) + h}}$. The double radical $\sqrt{E + \sqrt{G}}$, $G \geq 0$, $E + \sqrt{G} \geq 0$ ($E \geq 0 \Rightarrow E + \sqrt{G} \geq 0$, while if $E < 0$ the radical $E + \sqrt{G} \geq 0 \iff G \geq E^2$) can be written in the form $\sqrt{E + \sqrt{G}} = M + \sqrt{N}$ if $E = M^2 + N$, $G = 4M^2N$. In our case $E = P_2$, $G = P_2^2 - P_4(\Phi) + h$ it means that one must find linear polynomial $M_1(\Phi)$ and quadratic one $N_2(\Phi)$ etc.

Remark 4.5. For sake of simplicity we shall define only strong and weak solutions of Eq. (4.2). Similar definitions are given for the generalized Camassa-Holm equation Eq. (4.1).

Definition 4.1. A strong solution of Eq. (4.2) is called a function $u \in C([0,T] : H_x^3) \cap C^1([0,T] : H_x^2)$ that satisfies the Cauchy problem to Eq. (4.2) in the space $C([0,T] : L_x^2)$.

Definition 4.2. Let the function $u_0 \in H_x^1$ be given. A function $u \in [0,T] \times \mathbf{R}_x^1$ is called a weak solution to the Cauchy problem (1), $u|_{t=0} = u_0(x)$ if $u \in L_2([0,T] : H_x^1)$ and satisfies the integral identity

$$\int_0^T \int_{\mathbf{R}_x^1} (u\varphi_t + F(u)\varphi_x)dxdt + \int_{\mathbf{R}_x^1} u_0(x)\varphi(x,0)dx = 0$$

for all $\varphi \in C_0^\infty(\mathbf{R}^2)$, $supp\ \varphi \subset \{(-T,T) \times \mathbf{R}_x^1\}$.

The nonlinear operator $F(v) = (\frac{v^2}{2} + p * (v^2 + \frac{1}{2}v_x^2))$, $p(x) = \frac{1}{2}e^{-|x|}$ is defined for each $v \in H^1(\mathbf{R})$, i.e. v, $v_x \in L_{2,loc}$, while $p * w$ is convolution of p and w.

One can prove that every strong solution is a weak solution. Moreover, if the weak solution $u \in C([0,T] : H^3) \cap C^1([0,T] : H^2)$ then it is a strong solution.

Surprisingly, there are no nontrivial traveling waves for Eq. (4.2) which are strong solutions defined in [36].

Chapter 5

Introduction to the dressing method and application to the cubic NLS

5.1 Introduction

We propose below several historical notes on the dressing method used for finding of soliton type solutions of different equations of mathematical physics including the cubic Schrödinger operator (NLS). In the origins of that approach are the Swedish mathematician Bäcklund [15] and French mathematician G. Darboux [40] (see his investigations from around 1915). The interest to this method was renewed in the early 1970s of the last century by V. Zakharov and A. Shabat (see for example [145], [146]). Numerious papers were published in that direction, often jointly by Zakharov-Shabat and their collaborators S. Manakov, A. Mikhailov and many others ([144]). There is a big group of Bulgarian specialists in mathematical physics working in that domain and having good achievements there. Among them we shall mention only V. Gerdjikov and his scientific partners ([52], [71], [81], [132]). As the aim of our book is to acquaint the reader with several mathematical methods used for solving of different classes of nonlinear evolution PDEs we are not going to enter into historical details. We shall explain the point of the matter by finding soliton type solutions of the 1-D cubic nonlinear Schrödinger equation (NLS). Certainly, some difficulties will appear in that case and it will be shown (with some details) how to overcome them. To do this the corresponding mathematical machinery will be developed and explained. At the end of the chapter a more general class of Volterra type equations is studied and a large family of solutions of Kadomtsev-Petviashvili equation is constructed [144]. To simplify the things we shall separate our proof into several parts.

5.2   Preliminary notes

We shall consider the NLS

$$iq_t + q_{xx} + 2q|q|^2 = 0, \tag{5.1}$$

where $q : \mathbf{R}^2 \to \mathbf{C}^1$, $q = q(x,t)$, $lim_{|x|\to\infty}q(x,t) = 0$, $x \in \mathbf{R}^1$, $t \in \mathbf{R}^1$ and we shall construct a soliton type solution of Eq. (5.1) via the dressing method.

(a) At first we shall write Eq. (5.1) in the Lax form $[L(\lambda), A(\lambda)] = 0$, where $[,]$ is the commutator of the linear first order systems of linear ODE

$$L(\lambda) = i\partial_x + Q - \lambda\sigma_3, \sigma_3 = \begin{pmatrix} 1 & 0 \\ 0 & -1 \end{pmatrix}, \tag{5.2}$$

$$A(\lambda) = i\partial_t + A_0 + \lambda A_1 + \lambda^2 A_2 \equiv i\partial_t + V$$

and λ is a parameter. The 2×2 operator $Q = \begin{pmatrix} 0 & q \\ q^* & 0 \end{pmatrix}$, $q^* = \bar{q}$. In this chapter we follow the physical notations writing q^* instead of the complex conjugated $\bar{q}$ of q.

As $[L(\lambda), A(\lambda)]$ is a second order polynomial with respect to λ we must equalize to 0 the coefficients in front of λ^0, λ^1, λ^2 obtaining this way that $[L(\lambda), A(\lambda)] = 0$ is equivalent to

$$A_0 = \frac{i}{2}[\sigma_3, Q_x] + |q|^2\sigma_3, A_1 = 2Q, A_2 = -2\sigma_3.$$

(b) Below we formulate some necessary and well known facts from the scattering theory. Consider

$$L\psi = i\partial_x\psi + (Q - \lambda\sigma_3)\psi = 0, \tag{5.3}$$

where the 2×2 matrix ψ is fundamental solution to Eq. (5.3), i.e. $det\psi \neq 0$ everywhere. Having in mind that $tr(Q - \lambda\sigma_3) = 0$ we get that $det\psi(x,t,\lambda) \equiv det\psi(t,\lambda)$. If $[L, A]\psi = 0$ then $L(A\psi) = 0 \Rightarrow A\psi = \psi f(t,\lambda)$ (from the ODE theory). Therefore, $A\psi \equiv i\partial_t\psi + (A_0 + \lambda A_1 + \lambda^2 A_2)\psi = \psi(x,t,\lambda)f(t,\lambda)$. Put $V(t,x,\lambda) = A_0 + \lambda A_1 + \lambda^2 A_2$. Assuming that $\partial_t\psi \to_{|x|\to\infty} 0$, $\partial_x q \to_{|x|\to\infty} 0$ we get: $lim_{|x|\to\infty}A_0 = lim_{|x|\to\infty}A_1 = 0$.

Then we have $lim_{|x|\to\infty}V(x,t,\lambda) = -2\sigma_3\lambda^2$. Taking $lim_{|x|\to\infty}V = f(t,\lambda)$, i.e. $f = -2\lambda^2\sigma_3 \iff f = f(\lambda)$ we conclude that

$$i\partial_t\psi + V(x,t,\lambda)\psi = \psi f, f = -2\lambda^2\sigma_3.$$

Definition 5.1. For $\lambda \in \mathbf{R}^1$ the fundamental solution (FS) $\psi_\pm$ of Eq. (5.3) is called Jost solution if

$$i\partial_x \psi_\pm + (Q - \lambda\sigma_3)\psi_\pm = 0, \qquad (5.4)$$

$lim_{x \to \pm\infty}\psi_\pm(x, t, \lambda)e^{i\lambda\sigma_3 x} = 1$, (1 being the unit matrix).

Thus, $det\psi_\pm \equiv 1$, as $det\ e^{-i\lambda\sigma_3 x} = 1$.

Put $\hat{\psi}_+ = \psi_+^{-1}$, $\lambda \in \mathbf{R}^1$ and define the scattering matrix $T(t, \lambda)$ by the formula

$$T(t, \lambda) = \hat{\psi}_+ \psi_- \iff \psi_+ T = \psi_-; det\ T = 1. \qquad (5.5)$$

From Eq. (5.5) it follows that

$$i\partial_t(\psi_+ T) + V(\psi_+ T) = \psi_+ T f(\lambda). \qquad (5.6)$$

On the other hand, $\psi_+ \sim e^{-i\lambda\sigma_3 x}$ for $x \to \infty$.

For $x \to +\infty$ Eq. (5.6) implies that

$$i\partial_t T + f(\lambda)T = T f(\lambda),$$

i.e. the scattering matrix T satisfies the linear ODE

$$i\partial_t T + [f, T] = 0. \qquad (5.7)$$

Therefore,

$$T(t, \lambda) = e^{if(\lambda)t}T(0, \lambda)e^{-if(\lambda)t}. \qquad (5.8)$$

Introduce now the auxiliary function

$$\xi_\pm = \psi_\pm e^{i\lambda\sigma_3 x} \Rightarrow lim_{|x| \to \infty}\xi_\pm(x, \lambda) = 1. \qquad (5.9)$$

One can easily see that $\xi_\pm(x, \lambda)$ is a solution of the following Volterra type linear integral equation:

$$\xi_\pm(x, \lambda) = 1 + i \int_{\pm\infty}^{x} e^{i\lambda\sigma_3(y-x)}Q\xi_\pm e^{-i\lambda\sigma_3(y-x)}dy, \qquad (5.10)$$

$$det\xi_\pm = det\psi_\pm det\ e^{i\lambda\sigma_3 x} = 1.$$

Exercise 1. Suppose that ψ is a solution of the Cauchy problem

$$i\partial_x\psi - A\psi = -Q\psi = g, A = \lambda\sigma_3$$

$$\psi(x_0) = y_0, e^{iAx_0}y_0 = 1.$$

Then prove that $\psi(x) = e^{-iAx} - i\int_{x_0}^{x} e^{iA(y-x)}g(y)dy$ and respectively $\xi(x) = \psi(x)e^{iAx}$ satisfies the integral equation

$$\xi(x) = 1 - i\int_{x_0}^{x} e^{iA(y-x)}g(y)e^{iAx}dy =$$

$$1 + i \int_{x_0}^{x} e^{iA(y-x)} Q(y) \psi(y) e^{iAx} dy =$$

$$1 + i \int_{x_0}^{x} e^{iA(y-x)} Q(y) \xi(y) e^{-iA(y-x)} dy.$$

For $x_0 = \pm\infty$, $\xi = \xi_\pm$ you will get Eq. (5.10).

Exercise 2. The functions ψ and A are defined in Exercise 1. Prove that $\xi = \psi(x) e^{iAx}$ satisfies the ODE

$$i\xi' + \lambda[\xi, \sigma_3] + Q\xi = 0. \tag{5.11}$$

Therefore, $i\xi'_\pm + \lambda[\xi_\pm, \sigma_3] + Q\xi_\pm = 0$, $\xi_\pm(\pm\infty) = \mathbf{1}$.

Exercise 3. Let $\psi(x)$ be a fundamental solution of the ODE $i\psi' + A(x)\psi = 0$ and $\hat\psi = \psi^{-1}$. Prove that $\hat\psi$ satisfies the ODE $i\hat\psi' - \hat\psi A = 0$.

5.3 Dressing method and Riemann-Hilbert problem. Short survey

Consider the linear 2×2 system of ODE

$$L_\lambda \kappa = (i\partial_x + Q(x) - \lambda\sigma_3)\kappa(x, \lambda) = 0. \tag{5.12}$$

To fix the ideas we shall assume that Q is antidiagonal 2×2 matrix (see Eq. (5.2)) whose elements belong to the Schwartz class of rapidly decreasing at infinity functions. If κ is a FS of Eq. (5.12) then $\det\kappa \neq 0$ and $\det\kappa(x, \lambda)$ does not depend on x, as $tr(Q - \lambda\sigma_3) = 0$. If Φ and Ψ are fundamental solutions of Eq. (5.12) then $\Phi = \Psi T(\lambda)$, $\lambda \in \mathbf{R}^1$, where the matrix

$$T(\lambda) = \begin{pmatrix} a^+(\lambda) & -b^-(\lambda) \\ b^+(\lambda) & a^-(\lambda) \end{pmatrix}. \tag{5.13}$$

If Φ, Ψ are Jost solutions of Eq. (5.12) then $\det\Psi = \det\Phi = 1$ and therefore $\det T(\lambda) = 1 \Rightarrow a^+(\lambda)a^-(\lambda) + b^+(\lambda)b^-(\lambda) = 1$, $\lambda \in \mathbf{R}^1$.

Define now the functions associated with Eq. (5.12):

$$\xi(x, \lambda) = \Psi(x, \lambda) e^{i\lambda\sigma_3 x} \Rightarrow \xi(x, \lambda) \to_{x\to\infty} 1 \tag{5.14}$$

$$\hat\psi(x, \lambda) = \Psi^{-1}(x, \lambda)$$

$$\varphi(x, \lambda) = \Phi(x, \lambda) e^{i\lambda\sigma_3 x} \Rightarrow \varphi(x, \lambda) \to_{x\to-\infty} 1$$

As we know from Exercises 2, 3

$$i\partial_x \xi + Q\xi - \lambda[\sigma_3, \xi] = 0$$

$$i\partial_x \varphi + Q\varphi - \lambda[\sigma_3, \varphi] = 0$$

$$i\partial_x\hat{\psi} - \hat{\psi}(Q - \lambda\sigma_3) = 0.$$

Moreover, from Eq. (5.10) we have

$$\left| \begin{array}{l} \xi(x,\lambda) = 1 + i\int_{+\infty}^{x} e^{i\lambda\sigma_3(y-x)}Q(y)\xi(y,\lambda)e^{-i\lambda\sigma_3(y-x)}dy \\ \varphi(x,\lambda) = 1 + i\int_{-\infty}^{x} e^{i\lambda\sigma_3(y-x)}Q(y)\varphi(y,\lambda)e^{-i\lambda\sigma_3(y-x)}dy. \end{array} \right. \quad (5.15)$$

Write down $\xi = \begin{pmatrix} \xi_{11} & \xi_{12} \\ \xi_{21} & \xi_{22} \end{pmatrix} \Rightarrow Q\xi = \begin{pmatrix} q\xi_{21} & q\xi_{22} \\ q^*\xi_{11} & q^*\xi_{12} \end{pmatrix}$. Having in mind that

$e^{\sigma_3 t} = \begin{pmatrix} e^t & 0 \\ 0 & e^{-t} \end{pmatrix}$ we conclude that

$$\xi(x,\lambda) = 1 + i\int_{+\infty}^{x} \begin{pmatrix} q\xi_{21} & q\xi_{22}e^{-i2\lambda(x-y)} \\ q^*\xi_{11}e^{2i\lambda(x-y)} & q^*\xi_{12} \end{pmatrix} dy.$$

We represent the 2×2 matrices ξ and φ into column form

$$\xi = (\xi^-, \xi^+), \xi^- = \begin{pmatrix} \xi_{11} \\ \xi_{12} \end{pmatrix}, \xi^+ = \begin{pmatrix} \xi_{12} \\ \xi_{22} \end{pmatrix}, \varphi = (\varphi^+, \varphi^-). \quad (5.16)$$

An analysis of the Volterra equations Eq. (5.15) shows that $\xi^+(\varphi^+)$ is analytic with respect to $\lambda \in \mathbf{C}^1$ for $\lambda \in \mathbf{C}_+^1 = \{Im\lambda > 0\}$, while $\xi^-(\varphi^-)$ is analytic with respect to λ for $\lambda \in \mathbf{C}_-^1 = \{Im\lambda < 0\}$. Moreover, for each $p \in \mathbf{N}$ there exists $\frac{d^p}{d\lambda^p}\xi^\pm(x,\lambda)$, $\frac{d^p\varphi^\pm}{d\lambda^p}(x,\lambda)$ being analytic in the domains of analyticity of $\xi^\pm(\varphi^\pm)$. (See [52].)

Remark 5.1. If the reader is interested in the mathematical proofs we give a short sketch of the proof of the analyticity of the Jost solutions in the upper and lower half-planes of $\mathbf{C}^1$. One can easily see that the vector-columns of $\xi^\pm$, $\varphi^\pm$ satisfy the following linear Volterra type integral equations:

$$\xi^-(x,\lambda) = \begin{pmatrix} 1 \\ 0 \end{pmatrix} + i\int_{+\infty}^{x} G_2(x-y,\lambda)Q(y)\xi^-(y,\lambda)dy \quad (5.17)$$

$$\xi^+(x,\lambda) = \begin{pmatrix} 0 \\ 1 \end{pmatrix} + \int_{+\infty}^{x} G_1(x-y,\lambda)Q(y)\xi^-(y,\lambda)dy$$

$$\varphi^+(x,\lambda) = \begin{pmatrix} 1 \\ 0 \end{pmatrix} + i\int_{-\infty}^{x} G_2(x-y,\lambda)Q(y)\varphi^-(y,\lambda)dy$$

$$\varphi^-(x,\lambda) = \begin{pmatrix} 0 \\ 1 \end{pmatrix} + i\int_{-\infty}^{x} G_1(x-y,\lambda)Q(y)\varphi^-(y,\lambda)dy,$$

where the corresponding Green functions are given by the formulas

$$G_1(x-y,\lambda) = \begin{pmatrix} e^{-i2\lambda(x-y)} & 0 \\ 0 & 1 \end{pmatrix},$$

$$G_2(x-y,\lambda) = \begin{pmatrix} 1 & 0 \\ 0 & e^{2i\lambda(x-y)} \end{pmatrix}.$$

As it is well known ([5], [52], [133], [135]) Volterra equations are solved by Neumann series. The convergence of that series in λ in $\mathbf{C}_\pm^1$ is ensured by the exponentials $e^{\pm 2i\lambda(x-y)}$. For example, in the equation satisfied by ξ^- we have that $\infty > y > x \Rightarrow |e^{2i\lambda(x-y)}| = e^{-2Im\lambda(x-y)}$. The latter exponent is decreasing for $Im\lambda < 0$, i.e. for $\lambda \in \mathbf{C}_-^1$.

One can also show that the elements of the scattering matrix $T(\lambda)$ belong to the Schwartz class in x for $\lambda \in \mathbf{R}^1$.

Definition 5.2. The functions $\kappa^+(\kappa^-)$ are fundamental analytic solutions (FAS) of Eq. (5.12) if they are obtained by a combination of two columns of Jost type solutions that are analytic in $\mathbf{C}_+^1(\mathbf{C}_-^1)$:

$$\kappa^+(x,\lambda) = (\varphi^+, \xi^+)e^{-i\lambda\sigma_3 x} = (\Phi^+, \Psi^+) \qquad (5.18)$$
$$\kappa^-(x,\lambda) = (\xi^-, \varphi^-)e^{-i\lambda\sigma_3 x} = (\Psi^-, \Phi^-).$$

As we know, each two fundamental solutions of Eq. (5.12) can be compared. Thus, we can compare $\kappa^\pm$ with Φ, Ψ. Routine calculations show that

$$\kappa^+(x,\lambda) = \Psi(x,\lambda)\begin{pmatrix} a^+ & 0 \\ b^+ & 1 \end{pmatrix} = \Phi(x,\lambda)\begin{pmatrix} 1 & b^- \\ 0 & a^+ \end{pmatrix}, \qquad (5.19)$$

$$\kappa^-(x,\lambda) = \Psi(x,\lambda)\begin{pmatrix} 1 & -b^- \\ 0 & a^- \end{pmatrix} = \Phi(x,\lambda)\begin{pmatrix} a^- & 0 \\ -b^+ & 1 \end{pmatrix}$$

(see Eq. (5.13) for the link between Φ and Ψ: $\Phi = \Psi T(\lambda)$). According to Eq. (5.19): $det\kappa^+ = a^+$, $det\kappa^- = a^-$. Similar considerations are valid for $\hat\Psi(x,\lambda)$. In order to find a link between the FAS $\kappa^\pm$ we use Eq. (5.19) and find that for $\lambda \in \mathbf{R}^1$.

$$\kappa^+ = \Psi(x,\lambda)\begin{pmatrix} a^+ & 0 \\ b^+ & 1 \end{pmatrix} = \kappa^-\begin{pmatrix} 1 & -b^- \\ 0 & a^- \end{pmatrix}^{-1}\begin{pmatrix} a^+ & 0 \\ b^+ & 1 \end{pmatrix} = \kappa^- G_0(\lambda),$$

where $G_0(\lambda) = \frac{1}{a^-(\lambda)}\begin{pmatrix} 1 & b^-(\lambda) \\ b^+(\lambda) & 1 \end{pmatrix}$.

Conclusion: The FAS $\kappa^\pm$ of Eq. (5.12) are analytic with respect to λ in $\mathbf{C}_+^1$, respectively in $\mathbf{C}_-^1$, while for each $\lambda \in \mathbf{R}^1$ we have

$$\kappa^+(x,\lambda) = \kappa^-(x,\lambda)G_0(\lambda). \qquad (5.20)$$

Unfortunately, $\kappa^\pm$ are not normalized at infinity, as

$$\kappa^+ \sim e^{-i\lambda\sigma_3 x}\begin{pmatrix} a^+ & 0 \\ b^+ & 1 \end{pmatrix}, x \to \infty,$$

$$\kappa^+ \sim e^{-i\lambda\sigma_3 x}\begin{pmatrix} 1 & b^- \\ 0 & a^+ \end{pmatrix}, x \to -\infty,$$

$$\kappa^- \sim e^{-i\lambda\sigma_3 x}\begin{pmatrix} 1 & -b^- \\ 0 & a^- \end{pmatrix}, x \to -\infty.$$

In order to normalize these functions we introduce the auxiliary function

$$\eta^\pm(x,\lambda) = \kappa^\pm(x,\lambda)e^{i\lambda\sigma_3 x}. \tag{5.21}$$

Taking into account Eq. (5.20) we get

$$\eta^+(x,\lambda) = \eta^-(x,\lambda)G(x,\lambda), \tag{5.22}$$

where $G(x,\lambda) = e^{-i\lambda\sigma_3 x}G_0(\lambda)e^{i\lambda\sigma_3 x}$, $\lambda \in \mathbf{R}^1$.

Later on we shall prove that we can take $lim_{\lambda\to\infty}\eta^\pm(x,\lambda) = \mathbf{1}$, concluding this way that $\eta^\pm(x,\lambda)$ is normalized at infinity: $\lambda \to \infty \Rightarrow \eta^\pm(x,\lambda) \to \mathbf{1}$.

The functions G_0, G are called sewing functions of Eq. (5.20), Eq. (5.22), $\lambda \in \mathbf{R}^1$.

The fact that φ, ψ are FS of Eq. (5.12) implies that $\kappa^\pm$ are FAS of Eq. (5.12), i.e. $\eta^\pm$ according to Exercise 2 satisfy the system

$$i\frac{d\eta^\pm}{dx} + Q(x)\eta^\pm - \lambda[\sigma_3, \eta^\pm] = 0. \tag{5.23}$$

As $\eta^\pm(x,\lambda)$ are analytic for $\lambda \in \mathbf{C}_\pm^1$ they can be developed about $\lambda = \infty$ at the series

$$\eta^\pm(x,\lambda) = \eta_0^\pm(x) + \sum_{k=1}^\infty \eta_k^\pm(x)\lambda^{-k}, \lambda \in \mathbf{C}_\pm^1. \tag{5.24}$$

Substituting Eq. (5.24) in Eq. (5.23) we obtain:
The coefficient in front of λ^1 : $[\sigma_3, \eta_0^\pm] = 0 \Rightarrow \eta_0^\pm$ is a diagonal matrix;
the coefficient in front of λ^0 : $i\frac{d}{dx}\eta_0^\pm(x) + Q\eta_0^\pm - [\sigma_3, \eta_1^\pm] = 0$;
the coefficient in front of λ^{-1} : $i\frac{d}{dx}\eta_1^\pm + Q\eta_1^\pm - [\sigma_3, \eta_2^\pm] = 0$;

....

the coefficient in front of λ^{-k} : $i\frac{d}{dx}\eta_k^\pm + Q\eta_k^\pm - [\sigma_3, \eta_{k+1}^\pm] = 0$;

....

It is reasonable to split $\eta_k^\pm$ into diagonal and antidiagonal (off-diagonal) parts, namely

$$\eta_k^\pm = (\eta_k^\pm)^d + (\eta_k^\pm)^f$$

as $[\sigma_3, \eta_k^\pm]$ is an off-diagonal matrix and the product of diagonal and off-diagonal matrix is an off-diagonal one.

One can easily check that $i\frac{d}{dx}\eta_0^{\pm} = 0 \Rightarrow \eta_0^{\pm}(x) = const$. Without loss of generality we take $\eta_0^{\pm} = \mathbf{1}$.

Therefore,

$$Q - [\sigma_3, \eta_1^{\pm}] = 0. \tag{5.25}$$

Splitting Eq. (5.25) into diagonal and antidiagonal parts and having in mind that Q is off-diagonal and σ_3 is diagonal matrix we conclude that

$$(\eta_1^+)^f = (\eta_1^-)^f = \frac{1}{4}[\sigma_3, Q]. \tag{5.26}$$

On the other hand,

$$i\frac{d}{dx}(\eta_1^{\pm}(x))^d + Q(x)(\eta_1^{\pm}(x))^f = 0. \tag{5.27}$$

Combining Eq. (5.26), Eq. (5.27) we get:

$$\eta_1^+(x) = \frac{1}{2}\begin{pmatrix} -i\int_{-\infty}^x |q|^2 dy & q(x) \\ -q^*(x) & i\int_{+\infty}^x |q|^2 dy \end{pmatrix}$$

etc. From Eq. (5.21) and $\kappa^+ \to_{x\to\infty} e^{-i\lambda\sigma_3 x}\begin{pmatrix} a^+ & 0 \\ b^+ & 1 \end{pmatrix}$ we get that

$$a^+(\lambda) = 1 - \frac{i}{2\lambda}\int_{-\infty}^{\infty} |q|^2 dy + O\left(\frac{1}{|\lambda|^2}\right),$$

$$b^+(\lambda) = O\left(\frac{1}{|\lambda|}\right).$$

One can prove that each solution $\eta^{\pm}$ of Eq. (5.22) with sewing function G_0, such that $lim_{\lambda\to\infty}\eta^{\pm}(x,\lambda) = \mathbf{1}$ satisfies Eq. (5.23), the latter being equivalent to Eq. (5.12).

Exercise 4. ([52]) Prove the above formulated assertion.

Hint. Define the function $X^{\pm}(x,\lambda)$ by the formula

$$X^{\pm}(x,\lambda) = \left(i\frac{d\eta^{\pm}}{dx} - \lambda[\sigma_3, \eta^{\pm}]\right)\hat{\eta}^{\pm}(x,\lambda). \tag{5.28}$$

By using Eq. (5.22) prove that $X^+(x,\lambda) = X^-(x,\lambda)$ for $\lambda \in \mathbf{R}^1$. Having in mind that X^+ is analytic in $\mathbf{C}_+^1$, X^- is analytic in $\mathbf{C}_-^1$ we conclude that X^+, X^- can be extended to an analytic function in $\mathbf{C}^1$ (entire function with respect to λ). Due to Eq. (5.24),

$$lim_{\lambda\to\infty}g^{\pm}(x,\lambda) = lim_{\lambda\to\infty}\left(i\frac{d\eta^{\pm}}{dx} - \lambda[\sigma_3, \eta^{\pm}]\right)\hat{\eta}^{\pm} =$$

$$-[\sigma_3, \eta_1^{\pm}(x)] = -Q(x)$$

(see Eq. (5.25)). Therefore, $g^+(x,\lambda) + Q(x)$, $g^-(x,\lambda) + Q$ extends to an entire function in $\mathbf{C}^1_\lambda$ and $lim_{\lambda\to\infty}(g^+(x,\lambda)+Q) = lim_{\lambda\to\infty}(g^-(x,\lambda)+Q) = 0$. Consequently, $g^\pm + Q \equiv 0$. Then

$$X^+\eta^+ = \left(i\frac{d\eta^\pm}{dx} - \lambda[\sigma_3,\eta^\pm]\right) = -Q\eta^+, Im\ \lambda > 0. \qquad (5.29)$$

Combining Eq. (5.29) and Eq. (5.24) with $\eta_0^\pm = 1$ we get

$$Q(x) = lim_{\lambda\to\infty}\lambda(-\eta^\pm\hat\sigma_3\hat\eta^\pm + \sigma_3).$$

Let κ be FS of Eq. (5.12). We remind of the reader that the matrix Q is Hermitian self adjoint if $Q^\dagger = Q$, where $Q = (a_{ij})$ and $Q^\dagger = (a^\dagger_{ij}) = (a^*_{ji}) = (\bar a_{ji})$. Therefore, the matrix $Q = \begin{pmatrix} 0 & q \\ q^* & 0 \end{pmatrix}$ is Hermitian self adjoint. Evidently if κ is a FS of Eq. (5.12) then $i\partial_x\hat\kappa - \hat\kappa(Q - \lambda\sigma_3) = 0$ which implies that $i\partial_x\hat\kappa^\dagger + (Q - \lambda^*\sigma_3)\hat\kappa^\dagger = 0$, i.e. $\hat\kappa^\dagger$ is a FS of Eq. (5.12) but with λ^* instead of λ.

Conclusion. If $\kappa(x,\lambda)$ a FS of $L_\lambda\kappa = 0$ then $\tilde\kappa = \hat\kappa^\dagger(x,\lambda^*)$ is a FS of $L_{\lambda^*}\tilde\kappa = 0$.

Equation (5.20), Eq. (5.22) are important examples illustrating the so-called multiplicative Riemann-Hilbert problem. By using the Riemann-Hilbert problem we can solve several NLS type equations, including the cubic one. The multiplicative Riemann-Hilbert problem is a boundary-value problem from the theory of the analytic functions in $\mathbf{C}^1$. Its solvability is reduced to the investigation of systems of linear singular integral equations (see [56], [101], [114], [133]). Thus, let $F^\pm(\lambda)$, $F(\lambda)$ be $n \times n$ matrices depending on the parameter $\lambda \in \mathbf{C}^1$. The contour Γ splits $\mathbf{C}^1$ into two parts: interior Γ_- and exterior Γ_+ (see Fig. 5.1):

The function $F(\lambda)$ is sufficiently smooth on Γ and prescribed there.

Formulation of the Riemann-Hilbert problem (regular case).

Find $n \times n$ matrix-valued analytic functions $F^\pm(\lambda)$ in $\Gamma_\pm \setminus \Gamma$ such that

$$F^+(\lambda) = F^-(\lambda)F(\lambda), \forall\lambda \in \Gamma \qquad (5.30)$$

$$det\ F^+(\lambda)(det\ F^-(\lambda)) \neq 0 \ \text{ in } \ \Gamma_+ \setminus \Gamma(\Gamma_- \setminus \Gamma).$$

We will impose the additional condition $F^+(\lambda_0) = F^+_0$, $det\ F^+_0 \neq 0$ for the sake of uniqueness (say, $\lambda_0 \in \Gamma_+ \setminus \Gamma$).

Suppose that $\lambda_0 = \infty$, $F^+_0 = 1$, i.e. $lim_{\lambda\to\infty}F^+(\lambda) = 1$. Then it is well known (see [56], [101], [114], [133]) that the solvability of Eq. (5.30) is reduced to the solvability of the following system of singular integral equations on Γ with unknown functions $f(\lambda)$:

$$f(\lambda)(1 + F(\lambda))(1 - F(\lambda))^{-1} + 1 + \frac{1}{2\pi i}\int_\Gamma \frac{f(\mu)d\mu}{\mu - \lambda} = 0.$$

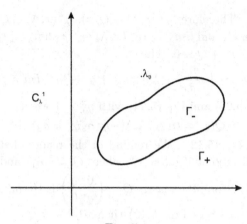

Fig. 5.1

The holomorphic functions $F^{\pm}(\lambda)$ are expressed by $\int_{\Gamma} \frac{f(\mu)d\mu}{\mu-\lambda}$ for $\lambda \in \Gamma_{\pm}$. As we mentioned before we shall describe the dressing method of Zakharov-Shabat by finding one soliton type solution of NLS Eq. (5.1). Suppose that q_0 is some solution of Eq. (5.1) and construct the potential $Q_0 = \begin{pmatrix} 0 & q_0 \\ q_0^* & 0 \end{pmatrix}$ and the operators L_0 and A_0 from Eq. (5.2). Consequently, we have the relations $Q_0 \to L_0$, $A_0 \to \psi_0$, $A_0(\lambda) = i\partial_t + A_0^{(0)}(\lambda) + \lambda A_1^{(0)}(\lambda) + \lambda^2 A_2^{(0)}$,

$$L_0\psi_0 = (i\partial_x + Q_0 - \lambda\sigma_3)\psi_0 = 0, det\psi_0 \neq 0. \qquad (5.31)$$

We are looking for a new function $g(x, t, \lambda)$ called a dressing factor and such that: $\psi_1 = g\psi_0$, $lim_{\lambda\to\infty}g(\lambda) = \mathbf{1}$ and

$$L_1\psi_1 = (i\partial_x + Q_1 - \lambda\sigma_3)\psi_1 = 0, \qquad (5.32)$$

where $Q_1 = \begin{pmatrix} 0 & q_1 \\ q_1^* & 0 \end{pmatrix}$. Briefly, $\psi_0 \to \psi_1 \to Q_1$.

Substituting ψ_1 with $g\psi_0$ in Eq. (5.32) we find that g is a solution of the ODE

$$i\partial_x g + Q_1 g - gQ_0 - \lambda[\sigma_3, g] = 0. \qquad (5.33)$$

From the second linear operator in Eq. (5.2) with $A(\lambda)$, denoted by A_1 we have that if $i\partial_t\psi_0 + V^{(0)}\psi_0 = 0$ then

$$A_1\psi_1 = (i\partial_t g - gV^{(0)} + V^{(1)}g)\psi_0, \qquad (5.34)$$

where $V^{(0)} = A_0^{(0)} + \lambda A_1^{(0)} + \lambda^2 A_2^{(0)}$, $V^{(1)} = A_0^{(1)} + \lambda A_1^{(1)} + \lambda^2 A_2^{(1)}$.

This is the Ansatz for the dressing factor g:

$$g(x,\lambda) = 1 + \frac{A(x)}{\lambda - \mu}, \lambda, \mu \in \mathbf{C}^1. \tag{5.35}$$

g must have a polar singularity because if g is an entire and bounded function then $g \equiv const$ according to the Liouville theorem.

In a similar way we are looking for

$$\hat{g}(x,\lambda) = 1 + \frac{B(x)}{\lambda - \nu}, \nu \in \mathbf{C}^1. \tag{5.36}$$

As we know $\tilde{\psi}_1 = \hat{\psi}_1^\dagger(x,\lambda^*)$ satisfies Eq. (5.32) with λ^* instead of λ and $\tilde{\psi}_0(x,\lambda^*)$ is a solution of Eq. (5.31) with λ^* instead of λ. Assuming that in both cases the dressing factor is the same we obtain that $\tilde{\psi}_1 = g\tilde{\psi}_0$, as $\psi_1 = g\psi_0$. One can easily see that $\tilde{\psi}_1 = g\tilde{\psi}_0 \Rightarrow \psi_1 = \hat{g}^\dagger \psi_0 \Rightarrow \hat{g}(x,\lambda^*) = g^\dagger(x,\lambda)$. Therefore, $A = B^\dagger, \mu = \nu^*$.

Letting $\lambda \to \infty$ in Eq. (5.33) that implies $lim_{\lambda \to \infty} \partial_x g = 0$, we get

$$Q_1(x) - Q_0(x) - [\sigma_3, A(x)] = 0. \tag{5.37}$$

If $A^\dagger = A$ then $[\sigma_3, A]$ has the form $\begin{pmatrix} 0 & r \\ r^* & 0 \end{pmatrix}$.

Our next step is to find $A(x)$. Having in mind that $g\hat{g} = 1$ and $\mu(\mu^*)$, $Im\mu \neq 0$ are poles of $g(\hat{g})$ we find

$$0 = lim_{\lambda \to \mu}(\lambda - \mu)g\hat{g} \Rightarrow A\left(1 + \frac{A^\dagger}{\mu - \mu^*}\right) = 0.$$

In a similar way $(1 + \frac{A}{\mu^* - \mu})A^\dagger = 0$.

Put $A = (\mu - \mu^*)\mathbf{P}(x) \Rightarrow A^\dagger = (\mu^* - \mu)\mathbf{P}^\dagger$. Evidently,

$$\mathbf{P}(1 - \mathbf{P}^\dagger) = 0, (1 - \mathbf{P})\mathbf{P}^\dagger = 0. \tag{5.38}$$

Consequently, $\mathbf{P} = \mathbf{PP}^\dagger = \mathbf{P}^\dagger$, i.e. $\mathbf{P}^2 = \mathbf{P}$ is a projector. As we know from the linear algebra $\mathbf{P}$ can be written in the form $\mathbf{P} = \frac{XF^T}{F^T X}$, where F^T is the transposed operator to F,

$$X = \begin{pmatrix} a_1(x) \\ a_2(x) \end{pmatrix}, F = \begin{pmatrix} b_1(x) \\ b_2(x) \end{pmatrix} \Rightarrow F^T = (b_1, b_2). \tag{5.39}$$

Thus, $XF^T = \begin{pmatrix} a_1b_1 & a_1b_2 \\ a_2b_1 & a_2b_2 \end{pmatrix}$, $detXF^T = 0$ but we assume that the rank $XF^T = 1$; $F^T X = a_1b_1 + a_2b_2$.

Later on we shall choose $X = F^*$ ($F = (F_{ij}) \iff F^* = (F_{ij}^*)$, $F^T = (F_{ji})$, $F^\dagger = (F_{ji}^*)$, F is Hermitian iff $F = F^\dagger$, $(F^T)^* = F^\dagger$).

Therefore,

$$X = \begin{pmatrix} a_1 \\ a_2 \end{pmatrix} \neq 0, F = \begin{pmatrix} a_1^* \\ a_2^* \end{pmatrix}, \mathbf{P} = \frac{1}{|a_1|^2 + |a_2|^2} \begin{pmatrix} |a_1|^2 & a_1 a_2^* \\ a_1^* a_2 & |a_2|^2 \end{pmatrix}. \quad (5.40)$$

Then

$$[\sigma_3, \mathbf{P}] = \frac{2}{|a_1|^2 + |a_2|^2} \begin{pmatrix} 0 & a_1 a_2^* \\ -a_1^* a_2 & 0 \end{pmatrix}.$$

From Eq. (5.33) we have that

$$0 = lim_{\lambda \to \mu}(\lambda - \mu)Q_1 = [(-i\partial_x A + AQ_0 - \mu A\sigma_3) + \mu\sigma_3 A](1 - \mathbf{P}).$$

Thus,

$$(-i\partial_x \mathbf{P} + \mathbf{P}Q_0 - \mu\mathbf{P}\sigma_3 + \mu\sigma_3 \mathbf{P})(1 - \mathbf{P}) = 0 \Rightarrow \quad (5.41)$$

$$(-i\partial_x \mathbf{P} + \mathbf{P}Q_0 - \mu\mathbf{P}\sigma_3)(1 - \mathbf{P}) = 0, \quad (5.42)$$

as $\mathbf{P}(1 - \mathbf{P}) = 0$.

By differentiation with respect to x we find $\partial_x \mathbf{P} = \partial_x(\frac{1}{F^T X}XF^T)$, $F^T X$ being scalar function.

Putting the expression for $\partial_x P$ in Eq. (5.42) after some calculations we get

$$[X(-i\partial_x F^T + F^T Q_0 - \mu F^T \sigma_3) - i\partial_x X F^T](1 - \mathbf{P}) = 0. \quad (5.43)$$

Equation (5.43) implies that the expression in the brackets should be proportional to $\mathbf{P}$. To simplify the considerations we shall impose stronger requirements, namely

$$i\partial_x F^T - F^T(Q_0 - \mu\sigma_3) = 0 \quad (5.44)$$

(F^T is a vector row).

$$i\partial_x X + (Q_0 - \mu^*\sigma_3)X = 0 \quad (5.45)$$

(X is a vector column).

One can immediately see that Eq. (5.45) implies that $-i\partial_x XF^T = F^T X(Q_0 - \mu^*\sigma_3)\mathbf{P}$, i.e. Eq. (5.44), Eq. (5.45) imply Eq. (5.43).

If $\hat{\psi}_0$ is a FS of $i\partial_x\hat{\psi}_0 - \hat{\psi}_0(Q_0 - \mu\sigma_3) = 0$ then $F^T = F_0^T\hat{\psi}_0(x, \mu) \Rightarrow F = \hat{\psi}_0^T F_0$. As we mentioned before, we take $X = F^* = \hat{\psi}_0^\dagger(x, \mu)F_0^*$. Assume that F^T satisfies Eq. (5.44). Then $Q_0 = Q_0^\dagger$ implies that $X = F^*$ is a solution of Eq. (5.45). Our last problem is to find F_0. According to Eq. (5.34) $A_1\psi_1 = 0 \Rightarrow i\partial_t g - gV^{(0)} + V^{(1)}g = 0$. Then

$$0 = lim_{\lambda \to \mu}(\lambda - \mu)V^{(1)} = lim_{\lambda \to \mu}(-i\partial_t g + gV^{(0)})\hat{g}. \quad (5.46)$$

In a similar way as above with the only change $x \to t$ we find

$$i\partial_t F^T - F^T V^{(0)}(x, t, \mu) = 0. \tag{5.47}$$

As $\psi_0 \sim e^{-i\mu\sigma_3 x}$, $x \to \infty$ we have that $\hat{\psi}_0 \sim e^{i\mu\sigma_3 x}$, $x \to \infty$. Thus, $F^T \sim F_0^T(t)e^{i\mu\sigma_3 x}$ and as we know $\lim_{x\to\infty} V^{(0)} = -2\mu^2\sigma_3$.

According to Eq. (5.47) for $x \to +\infty$

$$i\partial_t F_0^T + 2\mu^2 F_0^T \sigma_3 = 0, \tag{5.48}$$

i.e.

$$F_0^T = F_{00}^T e^{2i\mu^2\sigma_3 t}, F_{00}^T = const. \tag{5.49}$$

Conclusion: $F = \hat{\psi}_0^T(x, \mu)F_0$, $F_0 = e^{2i\mu^2\sigma_3 t}F_{00}$, $X = F^* = \hat{\psi}_0^\dagger F_0^* = \hat{\psi}_0^\dagger e^{-2i\mu^{*2}\sigma_3 t}F_{00}^*$.

We are ready now to construct a soliton type solution of the NLS equation (the so-called bright soliton). Taking $Q_0 \equiv 0$ we have from Eq. (5.31) that $\psi_0(x, \mu) = e^{-i\mu\sigma_3 x} \Rightarrow \hat{\psi}_0 = e^{i\mu\sigma_3 x}$. Therefore,

$$F^* = X = e^{-i\mu^*\sigma_3 x}e^{-2i\mu^{*2}\sigma_3 t}F_{00}^* = e^{-i\mu^*(x+2\mu^* t)\sigma_3}F_{00}^*,$$

$F_{00}^* = \begin{pmatrix} c_1^* \\ 1/c_1^* \end{pmatrix}$. Suppose that $\mu = \alpha - i\beta$, $\beta \neq 0$. Then $Q_1 = (\mu - \mu^*)[\sigma_3, \mathbf{P}]$ according to Eq. (5.37), i.e. $Q_1 = -2i\beta[\sigma_3, \mathbf{P}]$ and $e^{-i\mu^*(x+2\mu^* t)\sigma_3} = \begin{pmatrix} e^{-i\mu^*(x+2\mu^* t)} & 0 \\ 0 & e^{i\mu^*(x+2\mu^* t)} \end{pmatrix}$, while $\begin{pmatrix} a_1 \\ a_2 \end{pmatrix} = X = \begin{pmatrix} e^{-i\mu^*(x+2\mu^* t)}c_1^* \\ \frac{e^{i\mu^*(x+2\mu^* t)}}{c_1^*} \end{pmatrix}$, $c_1 = const \neq 0$.

From Eq. (5.40) we obtain that

$$\mathbf{P} = \frac{1}{|c_1|^2|e^{-i\mu^*(x+2\mu^* t)}|^2 + |c_1|^{-2}|e^{i\mu^*(x+2\mu^* t)}|^2} \begin{pmatrix} |a_1|^2 & a_1 a_2^* \\ a_1^* a_2 & |a_2|^2 \end{pmatrix},$$

where $a_1 = e^{-i\mu^*(x+2\mu^* t)}c_1^*$, $a_2 = e^{i\mu^*(x+2\mu^* t)}/c_1^* \Rightarrow a_1 a_2 = 1$, while $a_1 a_2^* = \frac{a_1}{a_1^*}$.

Then $Q_1 = \frac{-4i\beta}{|a_1|^2 + |a_2|^2} \begin{pmatrix} 0 & a_1 a_2^* \\ a_1^* a_2 & 0 \end{pmatrix}$, i.e. the corresponding solution of the NLS equation is

$$q_1 = -4i\beta \frac{a_1 a_2^*}{|a_1|^2 + |a_2|^2} = \frac{-4i\beta a_1}{a_1^*(|a_1|^2 + |a_2|^2)}.$$

Taking $|c_1| = e^D$ we obtain $|a_1|^2 + |a_2|^2 = 2ch[2\beta(x + 4\alpha t) + D]$, i.e.

$$q_1 = -2i\beta \frac{e^{-2i\alpha x - 4it(\alpha^2 - \beta^2)}}{ch\, 2[\beta(x + 4\alpha t) + D]} \frac{c_1^*}{c_1}, \tag{5.50}$$

$\frac{1}{ch\, z} = sech\, z$.

If $c_1 = i$ then $D = 0$ and $-2i\beta\frac{c_1^*}{c_1} = 2i\beta$.

On the other hand, $c_1 = \frac{\sqrt{2}}{2}(1 + i) \Rightarrow D = 0$ and $-2i\beta\frac{c_1^*}{c_1} = -2\beta$.

5.4   Geometrical interpretation of the soliton solutions

It turns out that the NLS equation naturally appears in the study of inextensible moving curves in $\mathbf{R}^3$. To do this some basic facts from the classical differential geometry of smooth curves will be reminded [22], [132]. Assume that $\gamma \subset \mathbf{R}^3$ is a regular curve, $\gamma \in C^k$, $k \geq 3$, parametrized by its arclength parameter s (natural parameter), i.e. $\gamma : \vec{r} = \vec{r}(s)$, $\vec{r}$ being the position vector.

We denote by $\vec{t}$ the unit tangent vector of γ, while $\vec{n}_1$, $\vec{n}_2$ are unit normal vectors to γ such that $\{\vec{t}, \vec{n}_1, \vec{n}_2\}$ form the Frenet-Serret type of moving frame, $\vec{n}_2 = \vec{t} \times \vec{n}_1$.

We know that the curve satisfies the following linear system in the case of classical Frenet frame $\{\vec{t}, \vec{n}, \vec{b}\}$, $\vec{b} = \vec{t} \times \vec{n}$.

$$\begin{pmatrix} \vec{t} \\ \vec{n} \\ \vec{b} \end{pmatrix}_s = \begin{pmatrix} 0 & \kappa & 0 \\ -\kappa & 0 & \tau \\ 0 & -\tau & 0 \end{pmatrix} \begin{pmatrix} \vec{t} \\ \vec{n} \\ \vec{b} \end{pmatrix},$$

$\kappa(s)$ being the curvature and $\tau(s)$ being the torsion of γ.

It is shown in [22] that instead of $\{\vec{t}, \vec{n}, \vec{b}\}$ one can apply another frame $\{\vec{t}, \vec{n}_1, \vec{n}_2\}$ (natural Frenet frame) in which the corresponding ODE takes the form

$$\begin{pmatrix} \vec{t} \\ \vec{n}_1 \\ \vec{n}_2 \end{pmatrix}_s = \begin{pmatrix} 0 & k_1 & k_2 \\ -k_1 & 0 & 0 \\ -k_2 & 0 & 0 \end{pmatrix} \begin{pmatrix} \vec{t} \\ \vec{n}_1 \\ \vec{n}_2 \end{pmatrix} \qquad (5.51)$$

There is a simple link among κ, τ, k_1, k_2. Thus,

$$k_1 = \kappa cos\Theta, k_2 = \kappa sin\Theta, \kappa^2 = k_1^2 + k_2^2, \tau = \Theta_s, \Theta = \angle(\vec{n}_1, \vec{n}). \qquad (5.52)$$

From Eq. (5.51) it follows that $\vec{t}_s = k_1 \vec{n}_1 + k_2 \vec{n}_2$.

Suppose now that the inextensible curve γ moves in $\mathbf{R}^3$ in such a way that $\{\vec{t}, \vec{n}_1, \vec{n}_2\}$ remains orthogonal during the motion. In other words $\{\vec{t}, \vec{n}_1, \vec{n}_2\}$ is moving as solid body. According to the classical mechanics [88], the evolution of $\{\vec{t}, \vec{n}_1, \vec{n}_2\}$ is given by the system of ODE

$$\begin{pmatrix} \vec{t} \\ \vec{n}_1 \\ \vec{n}_2 \end{pmatrix}_t = \begin{pmatrix} 0 & a_1 & a_2 \\ -a_1 & 0 & a_3 \\ -a_2 & -a_3 & 0 \end{pmatrix} \begin{pmatrix} \vec{t} \\ \vec{n}_1 \\ \vec{n}_2 \end{pmatrix}. \qquad (5.53)$$

Certainly, if γ moves in $\mathbf{R}^3$ sweeping a smooth surface there, its position vector $\vec{r} = \vec{r}(s,t)$, $\vec{t} = \partial_s \vec{r}$ and t is the time variable. "Inextensible" means that s, t are independent variables. Evidently,

$$\vec{t}_t = a_1 \vec{n}_1 + a_2 \vec{n}_2.$$

Having in mind that $\vec{t}_{st} = \vec{t}_{ts}$ we get:

$$\vec{t}_{st} = (k_{1t} - k_2 a_3)\vec{n}_1 + (k_{2t} + a_3 k_1)\vec{n}_2 - (a_1 k_1 + a_2 k_2)\vec{t}$$

$$\vec{t}_{ts} = -(a_1 k_1 + a_2 k_2)\vec{t} + a_{1s}\vec{n}_1 + a_{2s}\vec{n}_2.$$

Therefore,

$$k'_{1t} = k_2 a_3 + a'_{1s} \iff a'_{1s} = k'_{1t} - k_2 a_3 \tag{5.54}$$

$$k'_{2t} = -a_3 k_1 + a'_{2s} \iff a'_{2s} = k'_{2t} + a_3 k_1.$$

In a similar way $a'_{3s} = a_1 k_2 - a_2 k_1$.

By definition the velocity vector $\vec{v} = \partial_t \vec{r}(s,t) = u\vec{t} + v_1 \vec{n}_1 + v_2 \vec{n}_2$. As $\vec{r}_{ts} = \vec{r}_{st} = \vec{t}_t = a_1 \vec{n}_1 + a_2 \vec{n}_2$ and according to Eq. (5.51): $\vec{r}_{ts} = (u'_s - k_1 v_1 - k_2 v_2)\vec{t} + (uk_1 + v'_{1s})\vec{n}_1 + (uk_2 + v'_{2s})\vec{n}_2$ we have

$$u'_s = k_1 v_1 + k_2 v_2 \tag{5.55}$$

$$a_1 = uk_1 + v'_{1s}$$

$$a_2 = uk_2 + v'_{2s}.$$

We shall assume further on that $v_1 = k'_{1s}$, $v_2 = k'_{2s}$.
Then Eq. (5.55) implies

$$u_s = \frac{1}{2}\partial_s(k_1^2 + k_2^2) \Rightarrow u = \frac{1}{2}(k_1^2 + k_2^2)$$

$$a_1 = k_{1ss} + k_1 \frac{k_1^2 + k_2^2}{2}$$

$$a_2 = k_{2ss} + k_2 \frac{k_1^2 + k_2^2}{2}.$$

One can easily obtain from Eq. (5.54) that

$$k_{jt} = k_{jsss} + \frac{3}{2}(k_1^2 + k_2^2)k_{js}, j = 1, 2, \tag{5.56}$$

i.e. the curvatures $k_{1,2}$ satisfy a system of PDE of the type mKdV.

Suppose now that we have a pure binormal motion of the curve, i.e. $u = 0$. According to Eq. (5.55)

$$v_1 = -k_2, v_2 = k_1. \tag{5.57}$$

As Eq. (5.55) implies $a_1 = -k_{2s}$, $a_2 = k_{1s}$ we see that

$$a_{3s} = -\frac{1}{2}\partial_s(k_1^2 + k_2^2) \Rightarrow a_3 = -\frac{k_1^2 + k_2^2}{2}.$$

We introduce now the complex curvature function

$$U(s,t) = k_1 + ik_2.$$

Applying Eq. (5.54) we obtain

$$k_{1t} = a_{1s} + k_2 a_3 = -k_{2ss} - k_2 \frac{(k_1^2 + k_2^2)^2}{2}$$

$$k_{2t} = k_{1ss} + k_1 \frac{k_1^2 + k_2^2}{2}.$$

This way we come to the conclusion (see [132]) that

$$iU_t + U_{ss} + \frac{|U|^2}{2} U = 0. \tag{5.58}$$

The formulae Eq. (5.50), Eq. (5.52) give us $U = \kappa e^{i\Theta}$,

$$U = const\ sech[2(\beta(s + 4\alpha t) + D)]e^{-2i\alpha s - 4it(\alpha^2 - \beta^2)}$$

$$\kappa = const\ sech[2(\beta(s + 4\alpha t) + D], \Theta = -2\alpha s - 4t(\alpha^2 - \beta^2),$$

$\tau = \Theta_s = -2\alpha.$

Consequently, the curve γ : $\begin{matrix} \kappa = 4\beta sech(2\beta s) \\ \tau = -2\alpha \end{matrix}$ moves in $\mathbf{R}^3$ sweeping

a smooth surface. The result is similar to the method of characteristics when the $2D$ surface in $\mathbf{R}^3$ is weaved by a family of characteristics being transversal to some non characteristic space curve. In our case γ is a spiral (a helix type curve) with constant torsion and curvature $\kappa \to_{|s|\to\infty} 0$. Its loop is moving with velocity $-4\alpha = 2\tau$.

For the geometrical illustration see Fig. 5.2.

Exercise 5. Consider the inextensible curve γ with Frenet moving frame $\{\vec{t}, \vec{n}, \vec{b}\}$ and satisfying the system

$$\begin{pmatrix} \vec{t} \\ \vec{n} \\ \vec{b} \end{pmatrix}_t = \begin{pmatrix} 0 & a & b \\ -a & 0 & c \\ -b & -c & 0 \end{pmatrix} \begin{pmatrix} \vec{t} \\ \vec{n} \\ \vec{b} \end{pmatrix}, \vec{r} = \vec{r}(s,t), \frac{\partial \vec{r}}{\partial s} = \vec{t},$$

$\vec{v} = \frac{\partial \vec{r}}{\partial t} = u\vec{t} + v\vec{n} + w\vec{b}$. Suppose that the motion is purely binormal, i.e. $\vec{v} = w\vec{b}$.

a). Prove that κ and τ satisfy the system

$$\kappa_t + 2w_s\tau + w\tau_s = 0 \tag{5.59}$$

$$\tau_t = \left(\frac{w_{ss} - \tau^2 w}{\kappa}\right)_s + \kappa w_s.$$

b). Put $\kappa = const$ and prove that Eq. (5.59) takes the form

$$w = \tau^{-1/2} \tag{5.60}$$

$$\tau_t = \frac{1}{\kappa}[(\tau^{-1/2})_{ss} - \tau^{-3/2} + \kappa^2 \tau^{-1/2}]_s.$$

(This is the extended H. Dym equation.)

c). Transform Eq. (5.60) into the standard Dym equation

$$\tau_t = (\tau^{-1/2})_{sss}. \tag{5.61}$$

d). Find non-trivial solutions (torsions) of Eq. (5.61).

Hint to d). Look for traveling wave solutions of Eq. (5.61): $\tau = \varphi(x-ct)$, $c = const$, $\xi = x-ct$. Reduce Eq. (5.61) to an ODE with respect to $\varphi(\xi)$ and verify that $\tau = \varphi = \frac{1}{\lambda^2}$, $\xi + C_3 = \pm \int \frac{\lambda d\lambda}{\sqrt{C_2\lambda^3 + 2c\lambda + C_1\lambda^2}}$, C_3, C_2, $C_1 - const$, i.e. the latter integral defines an elliptic function and can be expressed by the Weierstrass functions ρ, ρ'.

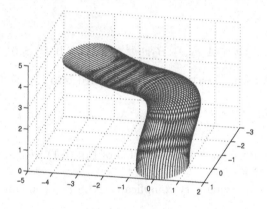

Fig. 5.2

In [53] the following system of 2 component nonlinear Schrödinger equations is considered:

$$i\frac{\partial q_1}{\partial t} + \frac{1}{2}\frac{\partial^2 q_1}{\partial x^2} + 2q_1(|q_1|^2 + 2|q_2|^2) + 2q_2^2 q_1^* = 0 \tag{5.62}$$

$$i\frac{\partial q_2}{\partial t} + \frac{1}{2}\frac{\partial^2 q_2}{\partial x^2} + 2q_2(2|q_1|^2 + |q_2|^2) + 2q_1^2 q_2^* = 0$$

It admits a Hamiltonian formulation with a Hamilton density given by

$$H = \frac{1}{2}\left(|\frac{\partial q_1}{\partial x}|^2 + |\frac{\partial q_2}{\partial x}|^2\right) - (|q_1|^2 + |q_2|^2)^2 - (q_1 q_2^* + q_1^* q_2)^2.$$

The first multicomponent NLS system is the Manakov one. Equation (5.62) were derived as a model for 4 wave mixing in the one-dimensional Bose-Einstein condensate by D. Zezyulin and V. Konotop.

Exercise 6. Find a soliton-type solution of Eq. (5.62).

Hint. After the linear change $q_1 = v_1 + v_2$, $q_2 = v_1 - v_2$ the system Eq. (5.62) is reduced to two scalar nonlinear cubic Schrödinger equations:

$$i\frac{\partial v_1}{\partial t} + \frac{1}{2}\frac{\partial^2 v_1}{\partial x^2} + 8v_1|v_1|^2 = 0 \tag{5.63}$$

$$i\frac{\partial v_2}{\partial t} + \frac{1}{2}\frac{\partial^2 v_2}{\partial x^2} + 8v_2|v_2|^2 = 0.$$

The change $v_{1,2} = \lambda r(\mu x, t) = \lambda \tilde{r}(x,t)$ with $\mu = \pm\sqrt{2}$, $\lambda = \frac{1}{2}$ transforms Eq. (5.63) into $i\tilde{r}_t + \tilde{r}_{xx} + 2|\tilde{r}|^2\tilde{r} = 0$ etc.

Exercise 7. Consider the cubic NLS equations

$$iu_t + A_1 u_{xx} + B_1 u|u|^2 = 0, A_1 > 0, B_1 > 0 \tag{5.64}$$

and

$$iv_t + A_2 v_{xx} + B_2 v|v|^2 = 0, A_2 > 0, B_2 > 0 \tag{5.65}$$

and prove that Eq. (5.64) can be transformed into Eq. (5.65) via a linear change.

Hint. Consider the change $u = \lambda v(t, \mu x)$, $\lambda \neq 0$, $\mu \neq 0$, λ, μ real, $\lambda = \pm\sqrt{\frac{B_2}{B_1}}$, $\mu = \pm\sqrt{\frac{A_2}{A_1}}$.

Therefore, if we know some solution of Eq. (5.1) we can find a solution of Eq. (5.64) too.

In [53] the following very complicated NLS two-component system was investigated:

$$-i\frac{\partial q_1}{\partial t} = \frac{1}{2}\frac{\partial^2 q_1}{\partial x^2} + q_1(3|q_1|^2 + 4|q_2|^2) + q_2(2|q_1|^2 - |q_2|^2) + \tag{5.66}$$
$$q_1^2 q_2^* + 2q_1^* q_2^2$$
$$-i\frac{\partial q_2}{\partial t} = \frac{1}{2}\frac{\partial^2 q_2}{\partial x^2} + q_1(|q_1|^2 - 2|q_2|^2) + q_2(4|q_1|^2 + 3|q_2|^2) -$$
$$q_1^* q_2^2 + 2q_1^2 q_2^*.$$

Exercise 8. Find a soliton type solution of Eq. (5.66).

Hint. The change of variables $q_1 = (w^3 - 1)v_1 + (w - 1)v_2$, $q_2 = (w^2 - w)v_1 + (w^4 - w^2)v_2$, $w^5 = 1$, $w \neq 1$ reduces Eq. (5.66) to two scalar NLS equations of the type Eq. (5.64) satisfied by v_1 and v_2.

5.5   Concluding remarks

1. In this chapter we constructed the explicit solution Eq. (5.50) of the NLS Eq. (5.2), which is one-soliton solution. Therefore, it will be interesting to construct N-soliton $(N \geq 2)$ solution of the same equation. For the first time it is found in [32], and more precise formula is provided in [5]. We shall present these formulae here for completeness. In every case $u(x,t) = \frac{G}{F}$, F-real valued function, satisfies NLS

$$iu_t + u_{xx} + u|u|^2 = 0. \tag{5.67}$$

Case 1. $N = 1$ (one-soliton solution): $G = e^{\eta_1}$, $F = 1 + e^{\eta_1 + \eta_1^* + \varphi_{11}^*}$, where $\eta_1 = p_1 x - \Omega_1 t$, $\Omega_1 = -ip_1^2$, $e^{\varphi_{11}^*} = \frac{1}{2}(p_1 + p_1^*)^{-2}$, $p_1 \in \mathbf{C}^1 \setminus 0$. We denote $p_1 = \alpha - i\beta$, $\alpha \neq 0$.

Standard calculations show that then we have

$$u(x,t) = \sqrt{2}|\alpha|sech[\alpha(x + 4\beta t) - ln(2^{3/2}|\alpha|)]e^{-i\beta x + it(\alpha^2 - \beta^2)}. \tag{5.68}$$

Case 2. The $N \geq 2$ soliton solution of Eq. (5.67) looks rather complicated:

$$F = \sum_{\bar{\mu}=0,1} D_1(\bar{\mu})exp\left(\sum_{i=1}^{2N}\mu_i\eta_i + \sum_{1\leq i<j\leq 2N}\varphi_{ij}\mu_i\mu_j\right)$$

$$G = \sum_{\bar{\mu}=0,1} D_2(\bar{\mu})exp\left(\sum_{i=1}^{2N}\mu_i\eta_i + \sum_{1\leq i<j\leq 2N}\varphi_{ij}\mu_i\mu_j\right),$$

where $\eta_i = p_i x - \Omega_i t$, $p_{i+N} = p_i^*$, $\Omega_{i+N} = \Omega_i^*$, $i = 1, 2, \ldots, N$, $\eta_{i+N} = \eta_i^*$, $\Omega_j = -ip_j^2$,

$$e^{\varphi_{ij}} = \begin{cases} \frac{1}{2}(p_i + p_j)^{-2} & \text{for } i = 1, 2, \ldots, N; \quad j = N+1, \ldots, 2N \\ \frac{1}{2}(p_i - p_j)^{-2} & \text{for } i = N+1, \ldots, 2N; j = N+1, \ldots, 2N. \end{cases}$$

Finally,

$$D_1(\bar{\mu}) = \begin{cases} 1 \text{ for } & \sum_{i=1}^{N}\mu_i = \sum_{i=1}^{N}\mu_{i+N} \\ 0 \text{ otherwise} \end{cases}$$

$$D_2(\bar{\mu}) = \begin{cases} 1 \text{ for } & 1 + \sum_{i=1}^{N}\mu_{i+N} = \sum_{i=1}^{N}\mu_i \\ 0 \text{ otherwise} \end{cases}.$$

2. In the book [88] the following formula concerning the moving curve γ and the solution U of Eq. (5.58) is given. Thus, let the corresponding integral surface $\gamma_t(s)$ is written as $\kappa = 4\beta sech[2\beta(s + 4\alpha t)]$, $\tau = -2\alpha$,

$\alpha, \beta \neq 0$. Put $\xi = 2\beta(s + 4\alpha t)$, $\nu = -\frac{\alpha}{\beta}$, $h = \frac{1}{\beta(\nu^2+1)}$. Consider $\gamma(s)$ in Cartesian coordinates with unit vectors i, j, k and let $\{\vec{t}, \vec{n}, \vec{b}\}$ be the standard Frenet frame of the curve. If the position vector $\vec{r} = r(s,t)$, $\vec{r}_s = t$, $\vec{r}_t = v$ then $\vec{r} = xi + yj + zk$ and $\gamma : r(s,0)$ takes the form:

$$x = -2hsechs\ sin\nu s$$
$$y = 2hsechs\ cos\nu s$$
$$z = s - htghs.$$

Certainly, the projection of γ on Oxy is bounded, while $z \sim s$ at $\pm\infty$.

5.6 Appendix. Volterra integral equations in infinite intervals

To begin with we shall consider the following example.
  Example 1. Consider the equation

$$\varphi(x) = \int_{-\infty}^{x} \frac{K(x)}{K(t)} \varphi(t)dt + f(x), \varphi \in C^0, f \in C^0, K(x) \in C^0, K(x) \neq 0 \tag{5.69}$$

everywhere.
  No conditions on K are imposed on $+\infty(-\infty)$. Equation (5.69) takes the form

$$\frac{\varphi(x)}{K(x)} = \frac{f(x)}{K(x)} + \int_{-\infty}^{x} \frac{\varphi(t)}{K(t)} dt; \tag{5.70}$$

Put $\psi(x) = \frac{\varphi(x)}{K(x)}$, $f_1(x) = \frac{f(x)}{K(x)} \Rightarrow$

$$\psi(x) = f_1(x) + \int_{-\infty}^{x} \psi(t)dt. \tag{5.71}$$

Assume that $\psi(t) \in L_1(-\infty, A)$, $\forall A = const$, $f_1(-\infty) = 0$.
  Then from Eq. (5.71) it follows that $\exists lim_{x \to -\infty} \psi(x) = 0$ and

$$\psi'(x) = f_1'(x) + \psi(x). \tag{5.72}$$

Formally solving the linear Eq. (5.72) we have:

$$\psi(x) = Ce^x + e^x \int_{-\infty}^{x} f_1'(t)e^{-t}dt = Ce^x + \int_{-\infty}^{x} f_1'(t)e^{-t+x}dt, C = const. \tag{5.73}$$

The formula is valid if $f_1'(t)e^{-t} \in L_1(-\infty, A)$, $\forall A$ and therefore $\int_{-N}^{x} f_1'(t)e^{-t}dt = \int_{-N}^{x} e^{-t}df_1 = e^{-x}f_1(x) - e^N f_1(-N) + \int_{-N}^{x} f_1(t)e^{-t}dt$.

Letting $N \to \infty$ and assuming that $lim_{x \to -\infty} e^{-x} f_1(x) = 0$, $f_1(x) e^{-x} \in L_1(-\infty, A)$, $\forall A$ we conclude that

$$\psi(x) = Ce^x + f_1(x) + \int_{-\infty}^x f_1(t) e^{x-t} dt \qquad (5.74)$$

satisfies Eq. (5.72) if $lim_{x \to -\infty} \frac{f(x)}{K(x)} = 0$, $lim_{x \to -\infty} e^{-x} \frac{f(x)}{K(x)} = 0$, $f_1(x) e^{-x} \in L_1(-\infty, A)$, $\forall A$. Evidently, $lim_{x \to -\infty} e^{-x} \frac{f(x)}{K(x)} = 0 \Rightarrow$ $lim_{x \to -\infty} \frac{f(x)}{K(x)} = 0$, $\frac{f(x)}{K(x)} \in L_1(-\infty, A)$.

Conclusion:

The ODE $\psi' = f_1'(x) + \psi$, $\psi(-\infty) = 0$ has the solution $\psi(x) = Ce^x + \frac{f(x)}{K(x)} + \int_{-\infty}^x \frac{f(t)}{K(t)} e^{x-t} dt$, $C = const$ if $f_1(x) e^{-x} \in L_1(-\infty, A)$, $\forall A$, $lim_{x \to -\infty} \frac{f(x)}{K(x)} e^{-x} = 0$.

Therefore, Eq. (5.69) is satisfied by

$$\varphi(x) = K(x)\psi(x) = CK(x)e^x + f(x) + \int_{-\infty}^x \frac{K(x)}{K(t)} e^{x-t} f(t) dt. \qquad (5.75)$$

Evidently, the kernel of Eq. (5.69) is one-dimensional while in the case $\varphi(x) = K(x)\psi(x) = CK(x)e^x + f(x) + \int_a^x \frac{K(x)}{K(t)} \varphi(t) dt$, $-\infty < a \le x \le b < \infty$ the solution is unique, that is its kernel is 0.

Below we study a more general example of Volterra type equation and construct a family of solutions of Kadomtsev-Petviashvili equation.

1. Volterra equations with degenerate kernels in infinite intervals.

Consider the equation

$$\varphi(x) = \sum_{k=1}^m \int_{-\infty}^x g_k(x) l_k(t) \varphi(t) dt + f(x), \, f, g_k, l_k \in C^0. \qquad (5.76)$$

Put $\int_{-\infty}^x l_k(t)\varphi(t) dt = \psi_k(x)$ and assume that $l_k \varphi \in L_1(-\infty, A)$, i.e. $\exists \psi_k(-\infty) = 0$.

Then Eq. (5.76) is reduced to the following linear system of ODE in the interval $(-\infty, A)$, $\forall A > 0$:

$$\begin{vmatrix} \psi_k'(x) = l_k(x) \sum_{j=1}^m g_j(x)\psi_j(x) + l_k f(x) \\ \psi_k(-\infty) = 0, k = 1, \dots, m \end{vmatrix} \qquad (5.77)$$

Solving Eq. (5.77) we get

$$\varphi(x) = \frac{\psi_k'(x)}{l_k(x)} = \sum_{j=1}^m g_j \psi_j + f(k = 1, \dots, m).$$

Asymptotic solutions of homogeneous systems of the type Eq. (5.77) can be found in the books of Codington-Levinson [35], Fedorjuk [45], Naimark and others.

Because of the lack of space we shall consider only the case $m = 2$ in Eq. (5.77):

$$\left| \begin{array}{l} \psi_1'(x) = l_1(x)(g_1(x)\psi_1(x) + g_2(x)\psi_2(x) + f(x)) \\ \psi_2'(x) = l_2(x)(g_1(x)\psi_1(x) + g_2(x)\psi_2(x) + f(x)) \end{array} \right. \tag{5.78}$$

$\psi_1(-\infty) = 0$, $\psi_2(-\infty) = 0$, $l_1 \neq l_2$, $l_1, l_2 \neq 0$, $g_2 \neq 0$.

Thus, $\psi_2(x) = \frac{\psi_1'}{l_1 g_2} = -\frac{g_1}{g_2}\psi_1 - \frac{f}{g_2} \Rightarrow \psi_2' = \frac{\psi_1'' l_1 g_2 - \psi_1'(l_1 g_2)'}{l_1^2 g_2^2} - \psi_1'\frac{g_1}{g_2} -$ $\psi_1(\frac{g_1}{g_2})' - (\frac{f}{g_2})' = \frac{l_2}{l_1}\psi_1'$. Consequently, Eq. (5.78) is equivalent to the scalar linear second order ODE:

$$\psi_1'' l_1 g_2 - \psi_1'((l_1 g_2)' + g_1 l_1^2 g_2 + l_2 l_1 g_2^2) - \psi_1(g_1' g_2 - g_2' g_1)l_1^2 = (f' g_2 - g_2' f)l_1^2, \tag{5.79}$$

$\psi_1(-\infty) = 0$.

As we know, the second order linear non homogeneous ODE of the type $\psi'' + g(x)\psi' + k(x)\psi = h(x)$ can be reduced by the change $\psi = ze^{-\frac{1}{2}\int g(x)dx}$ to the ODE

$$z'' + (k(x) - \frac{g^2(x)}{4} - \frac{g'(x)}{2})z = he^{\frac{1}{2}\int g(x)dx}. \tag{5.80}$$

Equations of the type Eq. (5.80) are easily studied by the Lagrange method of variation of constants. Moreover, the Wronskian of Eq. (5.80) is a constant $\neq 0$.

In the investigation of the homogeneous equation Eq. (5.80), i.e.

$$z'' - Q(x)z = 0, x \to -\infty \tag{5.81}$$

it is reasonable to apply the Liouville-Green approach (the so-called WKB method).

Proposition 5.1. [45]

Assume that $Q(x) \neq 0$ for $x \leq 0$, $\sqrt{Q}$ exists, $Re\sqrt{Q(x)} \geq 0$. Put $\delta(x, Q) = \frac{1}{8}\frac{Q''}{Q^{3/2}} - \frac{5}{32}\frac{(Q')^2}{Q^{5/2}}$, $\rho(x, Q) = |\int_a^x |\delta(t, Q)|dt|$, $a < 0$ and let $\rho(-\infty, Q) < \infty$.

(i) Then Eq. (5.81) possesses a solution $z_1(x)$ and such that $z_1 \sim Q^{-1/4}e^{\xi(x_0, x)}$, $x \to -\infty$, $\xi(x_0, x) = \int_{x_0}^x \sqrt{Q(t)}dt$, $x_0 < 0$.

(ii) Let $\frac{Q'}{Q^{3/2}} \to_{x \to -\infty} 0$. Then $z_1' \sim +Q^{1/4}e^{\xi(x_0, x)}$, $x \to -\infty$.

(iii) Equation (5.81) has a second solution z_2 which is linearly independent of z_1 and $z_2 \sim Q^{-1/4}e^{-\xi(x_0, x)}$, $x \to -\infty$.

If we are looking for solutions into explicit form the above described approach seems to be complicated from technical point of view. More precisely, if $Q(x) \geq 0$ is polynomial of even order $2s$, then $\xi(x_0, x) \sim -|x|^{s+1}$,

$x \to -\infty$ and we have a rapidly decreasing solution $z_2 \to 0$ of Eq. (5.81) for $x \to -\infty$. If we apply elementary tools it is reasonable to take in Eq. (5.78) $l_i = C_i e^{\lambda_i x}$, $i = 1, 2$, $g_i = D_i e^{\mu_i x}$, $i = 1, 2$, $f = C_5 e^{\nu x}$ and to look for $\psi_i = E_i e^{\gamma_i x}$.

2. Volterra integral equations in solving Kadomtsev-Petviashvili equation.

As it is shown in [144] the Kadomtsev-Petviashvili equation given by

$$\frac{\partial}{\partial x}(u_t - 6uu_x - u_{xxx}) = 3\alpha^2 u_{yy}, \alpha^2 < 0; \alpha = i \qquad (5.82)$$

can be solved by the dressing method (soliton solutions approach) following the below given procedure (see also [147]).

(a) Consider the system

$$i\frac{\partial F}{\partial y} + \frac{\partial^2 F}{\partial x^2} - \frac{\partial^2 F}{\partial z^2} = 0 \qquad (5.83)$$

$$\frac{\partial F}{\partial t} + 4\left(\frac{\partial^3 F}{\partial x^3} + \frac{\partial^3 F}{\partial z^3}\right) = 0$$

and find an appropriate solution F with separate variables.

(b) Solve the integral equation with respect to K with kernel F:

$$K(x, z, y, t) + F(x, z, y, t) + \int_{-\infty}^{x} K(x, z', y, t)F(z', z, y, t)dz' = 0. \quad (5.84)$$

(c) A solution of Eq. (5.82) is given by $u(x, y, t) = -2\frac{\partial}{\partial x}K(x, x, y, t)$.

To simplify the computations we shall further on omit the parameters (y, t) in Eq. (5.84). Thus, (a) we look for $F(x, z, y, t) = \varphi(x, y, t)\bar{\varphi}(z, y, t)$. Then the first equation of Eq. (5.83) takes the form

$$i\frac{\partial \varphi}{\partial y}\bar{\varphi} + \bar{\varphi}\varphi_{xx} + i\varphi\bar{\varphi}_y - \varphi\bar{\varphi}_{zz} = 0.$$

Take

$$i\frac{\partial \varphi}{\partial y} + \varphi_{xx} = 0 \Rightarrow i\bar{\varphi}_y - \bar{\varphi}_{zz} = 0 \Rightarrow i\varphi_y + \varphi_{zz} = 0. \qquad (5.85)$$

The latter equation holds due to Eq. (5.85) as $\varphi = \varphi(z, y, t)$ in it.

In a similar way one sees that

$$\frac{\partial \varphi}{\partial t} + 4\varphi_{xxx} = 0 \qquad (5.86)$$

implies the validity of the second equation of Eq. (5.83).

Thus, we suppose further on that

$$i\frac{\partial \varphi}{\partial y} + \varphi_{xx} = 0 \qquad (5.87)$$

$$\frac{\partial \varphi}{\partial t} + 4\varphi_{xxx} = 0.$$

Omitting (y, t), the Volterra Eq. (5.84) takes the form:

$$K(x, z) + F(x, z) + \int_{-\infty}^{x} K(x, z')F(z', z)dz' = 0, \qquad (5.88)$$

i.e.

$$K(x, z) + \varphi(x)\bar{\varphi}(z) + \int_{-\infty}^{x} \bar{\varphi}(z)\varphi(z')K(x, z')dz' = 0. \qquad (5.89)$$

It is reasonable to look for $K(x, z) = A(x)\bar{\varphi}(z)$, $A(x)$ being unknown. Therefore Eq. (5.89) can be rewritten as

$$A(x) + \varphi(x) + \int_{-\infty}^{x} A(x)|\varphi(z')|^2 dz' = 0, \qquad (5.90)$$

i.e.

$$A(x) = \frac{-\varphi(x)}{\int_{-\infty}^{x} |\varphi(z)|^2 dz + 1} \qquad (5.91)$$

and consequently

$$K(x, z, y, t) = \frac{-\varphi(x, y, t)\bar{\varphi}(z, y, t)}{1 + \int_{-\infty}^{x} |\varphi(z', y, t)|^2 dz'}. \qquad (5.92)$$

According to (c)

$$u(x, y, t) = 2\frac{\partial}{\partial x}\left(\frac{|\varphi|^2(x, y, t)}{1 + \int_{-\infty}^{x} |\varphi(z', y, t)|^2 dz'}\right) = 2\frac{\partial^2}{\partial x^2}ln\left(1 + \int_{-\infty}^{x} |\varphi(z, y, t)|^2 dz\right). \qquad (5.93)$$

Equation (5.93) is valid if $\varphi(x, y, t) \in L_2(-\infty, A)$, $\forall A$ with respect to x, and for fixed (y, t).

Let the parameter $k = k_1 + ik_2$ be such that $g = e^{Ax+By+Ct}$ verifies Eq. (5.87). The corresponding algebraic system is $iB + A^2 = 0$, $C + 4A^3 = 0$, i.e.

$$A = ik, B = -ik^2, C = 4ik^3. \qquad (5.94)$$

Supposing that $k_2 < 0 \Rightarrow |e^{Ax+By+Ct}| = |e^{ikx-ik^2y+4ik^3t}| = e^{-k_2x}|e^{-ik^2y+4ik^3t}|$, i.e. $g \in L_2(-\infty, A)$ with respect to x for fixed (y, t). Let $c(k_1, k_2)$ be compactly supported continuous function in $\{(k_1, k_2) : k_2 < 0\}$. Then

$$\varphi(x, y, t) = \int\int c(k_1, k_2)e^{ikx-ik^2y+4ik^3t}dk_1dk_2 \qquad (5.95)$$

exists and Eq. (5.93) is well defined.

Possible generalizations. Assume that the kernel F of Eq. (5.84) is degenerate and

$$F = \sum_{j=1}^{N} \varphi_j(x,y,t)\bar{\varphi}_j(z,y,t),$$

where φ_j are given by Eq. (5.95) for arbitrary $c_j \in C_0^0(k_2 < 0)$.

Then one can see that

$$u(x,y,t) = 2\frac{\partial^2}{\partial x^2}\ln\det A, \tag{5.96}$$

where $A = (A_{mn})_{m,n=1}^N$ and

$$A_{mn} = \delta_{nm} + \int_{-\infty}^{x} \varphi_n(z,y,t)\bar{\varphi}_m(z,y,t)dz.$$

As we are interested in explicit formulas to the solutions of Eq. (5.82), $\alpha = i$, we shall not give physical interpretation of Eq. (5.93), Eq. (5.96). The new moment is that the corresponding solutions depend on N-arbitrary functions c_j [144].

Chapter 6

Direct methods in soliton theory. Hirota's approach

6.1 Simplified Hirota's method in soliton theory

The famous Hirota's method was developed to find exact solutions of classes of evolution PDEs that can be written into a bilinear form. Unfortunately, finding these bilinear forms is a very difficult task. On the other hand, the approach itself is clear and non complicated. Following [59], [60], [136] we give a short sketch of a rather simple version of Hirota's method enabling us to construct solitary and soliton solutions of model examples of evolution PDEs. To do this logarithmic transformation is used. Here finding of bilinear forms is avoided and exact solutions are constructed via the perturbation method. We shall illustrate this approach by some equations of mathematical physics. Short description of the binary differential operator D and its applications is given in the second part of the Chapter (see also [3], [5], [63], [64], [65], [102]). Further development of that method is proposed in [96]. Soliton type solutions can be found by ISM only as it is done for example in [88].

We shall begin with the KdV equation

$$u_t + 6uu_x + u_{xxx} = 0. \qquad (6.1)$$

We make the logarithmic change

$$u = 2(ln\ f(x,t))_{xx} = 2\frac{ff_{xx} - f_x^2}{f^2}. \qquad (6.2)$$

Put $u = w_x$ in Eq. (6.1). Then Eq. (6.1) takes the form

$$w_{xt} + 6w_x w_{xx} + w_{xxxx} = 0 \Rightarrow w_t + 3w_x^2 + w_{xxx} = C(t). \qquad (6.3)$$

We take $C = 0$. Then

$$\frac{1}{2}w = \frac{f_x}{f} = (ln\ f)_x$$

(the corresponding additive constant is taken 0)

$$\frac{1}{2}u = \frac{ff_{xx} - f_x^2}{f^2} = \frac{f_{xx}}{f} - \frac{f_x^2}{f^2}$$

$$\frac{1}{2}u_x = \frac{f_{xxx}}{f} - 3\frac{f_x f_{xx}}{f^2} + 2\frac{f_x^3}{f^3}$$

$$\frac{1}{2}u_{xx} = \frac{f_{xxxx}}{f} - \frac{4f_x f_{xxx}}{f^2} - 3\frac{f_{xx}^2}{f^2} + 12\frac{f_x^2 f_{xx}}{f^3} - 6\frac{f_x^4}{f^4}$$

$$\frac{3}{2}u^2 = 6\frac{f_{xx}^2}{f^2} - 12\frac{f_{xx} f_x^2}{f^3} + 6\frac{f_x^4}{f^4}.$$

Therefore, $\frac{1}{2}(w_{xxx} + 3w_x^2) = \frac{1}{2}u_{xx} + \frac{3}{2}u^2 =$

$$\frac{f_{xxxx}}{f} - 4\frac{f_x f_{xxx}}{f^2} + 3\frac{f_{xx}^2}{f^2} = -\frac{1}{2}w_t = \frac{f_x f_t - f_{xt}f}{f^2}.$$

After the change Eq. (6.2) the KdV equation Eq. (6.1) is rewritten in the following form

$$f[f_{xxxx} + f_{xt}] - f_x f_t - 4f_x f_{xxx} + 3f_{xx}^2 = 0. \tag{6.4}$$

Denote by $\mathcal{L}(\partial_x, \partial_t) = \frac{\partial^2}{\partial x \partial t} + \frac{\partial^4}{\partial x^4}$ and by N the nonlinear (bilinear) differential operator

$$N(f,g) = -f_x g_t - 4f_x g_{xxx} + 3f_{xx}g_{xx}.$$

Conclusion: The equation Eq. (6.4), i.e. Eq. (6.1) after the change Eq. (6.2) takes the form

$$f\mathcal{L}(f) + N(f,f) = 0. \tag{6.5}$$

We are looking for a solution of Eq. (6.5) having the type (Ansatz)

$$f(x,t) = 1 + \sum_{n=1}^{\infty} \varepsilon^n f^{(n)}(x,t), \tag{6.6}$$

where $\varepsilon > 0$ is a parameter only (scaling) and not a small quantity. Substituting Eq. (6.6) in Eq. (6.5) and equate to 0 the coefficients standing in front of ε^j, $j = 1, 2, \ldots$ we get:

(a) $O(\varepsilon^1) : \mathcal{L}f^{(1)} = 0$
(b) $O(\varepsilon^2) : \mathcal{L}f^{(2)} = -N(f^{(1)}, f^{(1)})$
(c) $O(\varepsilon^3) : \mathcal{L}f^{(3)} = -f^{(1)}\mathcal{L}f^{(2)} - N(f^{(1)}, f^{(2)}) - N(f^{(2)}, f^{(1)})$ (6.7)

...

$O(\varepsilon^n) : \mathcal{L}f^{(n)} = -\sum_{j=1}^{n-1}[f^{(j)}\mathcal{L}(f^{(n-j)}) + N(f^{(j)}, f^{(n-j)})].$

In the Hirota's method the N-soliton solution of Eq. (6.1) is generated by $f^{(1)}$, the latter being a sum of exponents:

$$f^{(1)} = \sum_{i=1}^{N} e^{\Theta_i} = \sum_{i=1}^{N} e^{k_i x - w_i t + \delta_i}, \qquad (6.8)$$

k_i, w_i, δ_i being appropriate constants, $\Theta_i = k_i x - w_i t + \delta_i$.

Putting Eq. (6.8) in $\mathcal{L}f^{(1)} = 0$ we obtain $\mathcal{L}(k_i, -w_i) = 0$, $i = 1, \ldots, N$, where $\mathcal{L}(\partial_x, \partial_t)$ has the symbol $\mathcal{L}(k, w) = kw + w^4$, i.e. $\mathcal{L}(k_i, -w_i) = -k_i w_i + k_i^4 \iff k_i^3 = w_i$. The equality $k_i = w_i^3$ is called dispersion law. Compute now the right-hand side of Eq. (6.7) (b) for Eq. (6.8).

$$-N(f^{(1)}, f^{(1)}) = -\sum_{i,j=1}^{N} w_i k_j e^{\Theta_i + \Theta_j} + 4 \sum_{i,j=1}^{N} k_i k_j^3 e^{\Theta_i + \Theta_j} - \qquad (6.9)$$

$$3 \sum_{i,j=1}^{N} k_i^2 k_j^2 e^{\Theta_i + \Theta_j} = 3 \sum_{i,j}^{N} e^{\Theta_i + \Theta_j} k_i k_j^2 (k_j - k_i) =$$

$$3 \sum_{1 \leq i < j \leq N} k_i k_j (k_i - k_j)^2 e^{\Theta_i + \Theta_j},$$

as in the $sum(k_j^2 - k_i k_j) = k_i^2 + k_j^2 - k_i k_j - k_j k_i = (k_i - k_j)^2$ for $i < j$.

From Eq. (6.9) one can expect that

$$f^{(2)} = \sum_{1 \leq i < j \leq N} a_{ij} e^{\Theta_i + \Theta_j}, \qquad (6.10)$$

a_{ij} being unknown coefficients. Evidently,

$$\mathcal{L}f^{(2)} = \sum_{1 \leq i < j \leq N} [\mathcal{L}(k_i + k_j, -(w_i + w_j)) a_{ij} e^{\Theta_i + \Theta_j} = -N(f^{(1)}, f^{(1)}). \quad (6.11)$$

This way we conclude that

$$a_{ij} = \frac{3 k_i k_j (k_i - k_j)^2}{\mathcal{L}(k_i + k_j, -(w_i + w_j))} = \frac{3 k_i k_j (k_i - k_j)^2}{-(k_i^3 + k_j^3)(k_i + k_j) + (k_i + k_j)^4} = \qquad (6.12)$$

$$\frac{(k_i - k_j)^2}{(k_i + k_j)^2}, 1 \leq i < j \leq N.$$

We can continue in the same way with $f^{(3)}$. Having in mind that the computations are heavy it is better to stop here.

Conclusion: (a) Let $N = 1$. Then $f_1 = 1 + f^{(1)} = 1 + e^{\Theta}(\varepsilon = 1)$, $\Theta = kx - wt + \delta$, $w = k^3$.

(b) $N = 2$, $f_2 = 1 + e^{\Theta_1} + e^{\Theta_2} + a_{12} e^{\Theta_1 + \Theta_2}$, $a_{12} = \frac{(k_1 - k_2)^2}{(k_1 + k_2)^2}$.

One can show by computation that $f^{(n)} = 0$ for $n \geq 3$, $\varepsilon = 1$ in the expression for f_2.

(c) $N = 3$. Then

$$f_3 = 1 + e^{\Theta_1} + e^{\Theta_2} + e^{\Theta_3} + a_{12}e^{\Theta_1+\Theta_2} + a_{13}e^{\Theta_1+\Theta_3} + a_{23}e^{\Theta_2+\Theta_3} + b_{123}e^{\Theta_1+\Theta_2+\Theta_3},$$

where a_{ij} are given by Eq. (6.12) and $b_{123} = a_{12}a_{13}a_{23} = \frac{(k_1-k_2)^2(k_1-k_3)^2(k_2-k_3)^2}{(k_1+k_2)^2(k_1+k_3)^2(k_2+k_3)^2}$. Moreover, $f^{(n)} = 0$ for $n > 3$; $\varepsilon = 1$ in (c).

The N-soliton solution exists for every $N > 3$ and can be found in a similar way. To avoid technical difficulties it is better to prove the existence of N soliton solution of Eq. (6.1) inductively as in [5].

Remark 6.1. In the case (a) $u = 2(ln\, f_1)_{xx} = 2(ln(1 + e^{\Theta}))_{xx}$. After standard calculations one gets that

$$u = \frac{k^2}{2} sech^2 \frac{1}{2}(kx - k^3 t + \delta)$$

is a 1-soliton solution of Eq. (6.1). In fact, $2(ln(1 + e^{\Theta}))_{xx} = 2k^2 \frac{(e^{\Theta/2})^2}{(1+e^{\Theta})^2} = \frac{k^2}{2} sech^2 \Theta/2$.

Remark 6.2. The validity of (c) $N = 3$ can be proved by using algebraic identities as $(a + b + c)^3 - a^3 - b^3 - c^3 = 3(a + b)(b + c)(a + c)$ and others.

After some elementary algebra the two soliton solution of Eq. (6.1), i.e. (b), $N = 2$ is given by the formula

$$u = 2\frac{k_1^2 e^{\Theta_1} + k_2^2 e^{\Theta_2} + (k_1 - k_2)^2 e^{\Theta_1+\Theta_2}}{(1 + e^{\Theta_1} + e^{\Theta_2} + a_{12}e^{\Theta_1+\Theta_2})^2} +$$

$$2\frac{a_{12}e^{\Theta_1+\Theta_2}((k_1 + k_2)^2 + k_2^2 e^{\Theta_1} + k_1^2 e^{\Theta_2})}{(1 + e^{\Theta_1} + e^{\Theta_2} + a_{12}e^{\Theta_1+\Theta_2})^2}$$

where $\Theta_j = k_j x - k_j^3 t + \delta_j$, $j = 1, 2$ and $a_{12} = \frac{(k_1-k_2)^2}{(k_1+k_2)^2}$, $k_1 \neq \pm k_2$.

Concerning the interaction of two solitons, assume that $0 < k_1 < k_2$, $\delta_{1,2} = 0$. Put $\Theta_i = k_i \xi_i$, $i = 1, 2$, i.e. $\xi_i = x - k_i^2 t \Rightarrow \xi_2 = \xi_1 - (k_2^2 - k_1^2)t$. Fix ξ_1 and let $t \to \infty$. Therefore, $\Theta_2 \to -\infty \Rightarrow f_2 \sim 1 + e^{\Theta_1} \Rightarrow u = 2(ln\, f_2)_{xx} \sim \frac{k_1^2}{2} sech \frac{\Theta_1}{2}$. Assume now that for ξ_1 fixed $t \to -\infty$. Then $\Theta_2 \to \infty \Rightarrow f_2 \sim e^{\Theta_2}(1 + a_{12}e^{\Theta_1}) \Rightarrow u = 2(ln\, f_2)_{xx} \sim 2(ln e^{\Theta_2})_{xx} + 2(ln(1 + a_{12}e^{\Theta_1}))_{xx} = \frac{k_1^2}{2} sech \frac{ln\, a_{12} + \Theta_1}{2}$.

Therefore, the profile of the wave is conserved for $t \to +\infty$; the two solitons start at $t = -\infty$, then interact for finite T and propagating they reach $t = \infty$ but a phase change-translation equal to $ln\frac{(k_1-k_2)^2}{(k_1+k_2)^2} < 0$ appears. The qualitative picture of the interaction and propagation of the two solitons is given on Fig. 6.1.

In such elementary way as for KdV equation one can find 1, 2, 3 soliton solutions of the fourth order Boussineq equation (see [136]):

$$u_{tt} - u_{xx} - 3(u^2)_{xx} - u_{xxxx} = 0. \qquad (6.13)$$

Looking for a solution of the form $u = 2(ln\ f)_{xx}$ and repeating the procedure of Eq. (6.3) we get

$$f(f_{tt} - f_{xx} - f_{xxxx}) - (f_t^2 - f_x^2 - 4f_x f_{xxx} + 3f_{xx}^2) = 0. \qquad (6.14)$$

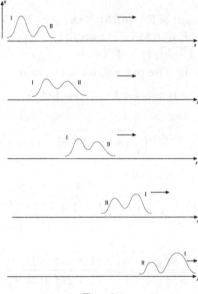

Fig. 6.1.

We obtained above exponential solutions of Eq. (6.5) via Eq. (6.6), where $f^{(1)}, \ldots, f^{(n)} \ldots$ were linear exponents, i.e. of the type $e^{Ax+Bt+C}$. It is possible to find polynomial solutions of Eq. (6.5) too. In fact, put $f^{(1)} = a_0 + a_1 x + a_2 x^2 + a_3 x^3 + bt$ in Eq. (6.6). Then $\mathcal{L}f^{(1)} = 0$ for arbitrary coefficients a_0, a_1, a_2, a_3, b. On the other hand, $N(f^{(1)}, f^{(1)}) = 0$ if $-(b + 24a_3)f_x^{(1)} + 3(4a_2^2 + 24a_2 a_3 x + 36a_3 x^2) \equiv 0$, i.e. for $b = 12a_3$, $a_2^2 = 3a_1 a_3$. Then we take $f^{(2)} = 0$, $\varepsilon = 1$ and $f_1 = 1 + a_0 + a_1 x + \sqrt{3a_1 a_2}x^2 + a_3(x^3 + 12t)$, $u(x,t) = 2(ln(1 + f^{(1)}))_{xx} = 2\frac{f_{1xx}(1+f^{(1)})-f_{1x}^2}{(1+f^{(1)})^2}$.

Taking $a_0 = a_1 = 0$, $a_3 = 1$ we find a rational solution $u = \frac{6x(2-(x^3-24t))}{(1+(x^3+12t))^2}$. The solution has a singularity at the semicubic parabola $x^3 + 12t = -1$.

Remark 6.3. If $u = 2(ln\ f)_{xx}$ then for each constants A, B we have $u = 2(ln(e^{Ax+Bt}f))_{xx}$.

Denote now $\mathcal{L} = \partial_{tt}^2 - \partial_{xx}^2 - \partial_{xxxx}^4$ and

$$N(f, g) = -(f_t g_t - f_x g_x - 4f_x g_{xxx} - 4f_x g_{xxx} + 3f_{xx} g_{xx}).$$

Conclusion: Equation (6.13) after the change Eq. (6.2) is written as

$$f\mathcal{L}f + N(f, f) = 0, \tag{6.15}$$

i.e. Eq. (6.5).

Looking for a solution of Eq. (6.15) of the type Eq. (6.6) we come to the Eq. (6.7). Similarly to Eq. (6.8) we obtain $\mathcal{L}(k_i, -w_i) = 0 \Rightarrow w_i^2 - k_i^2 - k_i^4 = 0 \iff w_i = -\varepsilon_i k_i \sqrt{1 + k_i^2}$, $\varepsilon_i^2 = 1$, i.e. $\Theta_i = k_i(x + \varepsilon_i \sqrt{1 + k_i^2}t) + \delta_i$. To fix the ideas let $\varepsilon_i = 1$. The one soliton solution $u = 2(ln(1 + e^{\Theta}))_{xx} = \frac{k_i^2}{2}sech^2 \frac{k_1(x+t\sqrt{1+k_1^2})}{2}$ as it was for Eq. (6.1). Let $N = 2$, $f^{(1)} = e^{\Theta_1} + e^{\Theta_2}$. Evidently,

$$-N(f^{(1)}, f^{(1)}) = 2k_1 k_2 e^{\Theta_1+\Theta_2}\left(\sqrt{1+k_1^2}\sqrt{1+k_2^2} - (1 + 2k_1^2 - 3k_1 k_2 + 2k_2^2)\right).$$

Taking $f^{(2)} = a_{12}e^{\Theta_1+\Theta_2}$ we have

$$\mathcal{L}(f^{(2)}) = a_{12}\mathcal{L}\left(k_1 + k_2, k_1\sqrt{1 + k_1^2} + k_2\sqrt{1 + k_2^2}\right)e^{\Theta_1+\Theta_2} = -N(f^{(1)}, f^{(1)}),$$

i.e.

$$a_{12} = \frac{\sqrt{1+k_1^2}\sqrt{1+k_2^2} - (1 + 2k_1^2 - 3k_1 k_2 + 2k_2^2)}{\sqrt{1+k_1^2}\sqrt{1+k_2^2} - (1 + 2k_1^2 + 3k_1 k_2 + 2k_2^2)}, \tag{6.16}$$

as $\mathcal{L}(k_1 + k_2, k_1\sqrt{1 + k_1^2} + k_2\sqrt{1 + k_2^2}) = (k_1\sqrt{1 + k_1^2} + k_2\sqrt{1 + k_2^2})^2 - (k_1 + k_2)^2 - (k_1 + k_2)^4 = 2k_1 k_2(\sqrt{1 + k_1^2}\sqrt{1 + k_2^2} - (1 + 2k_1^2 + 3k_1 k_2 + 2k_2^2))$.

We can see that $f^{(n)} = 0$ for $n \geq 3$ and therefore

$$f \equiv f_2 = 1 + e^{\Theta_1} + e^{\Theta_2} + a_{12}e^{\Theta_1+\Theta_2}, u = (ln\ f_2)_{xx}.$$

As it shown in [25] the 3-soliton solution is given by

$$f_3 \equiv f =$$

$$1 + e^{\Theta_1} + e^{\Theta_2} + e^{\Theta_3} + a_{12}e^{\Theta_1+\Theta_2} + a_{13}e^{\Theta_1+\Theta_2} + a_{23}e^{\Theta_2+\Theta_3} + b_{123}e^{\Theta_1+\Theta_2+\Theta_3}, \tag{6.17}$$

where

$$a_{ij} = \frac{\sqrt{1+k_i^2}\sqrt{1+k_j^2} - (2k_i^2 - 3k_ik_j + 2k_j^2 + 1)}{\sqrt{1+k_i^2}\sqrt{1+k_j^2} - (2k_i^2 + 3k_ik_j + 2k_j^2 + 1)}, i \neq j, \quad (6.18)$$

while

$$b_{123} = a_{12}a_{13}a_{23}, \quad (6.19)$$

i.e. the solution is the same as in (c), $N = 3$ for the KdV equation.

There are also explicit formulas for the $N = 4, 5$ soliton solution of Eq. (6.13) but they are very heavy. For example, for $N = 4$ $u = 2(ln\ f_4)_{xx}$, where

$$f_4 = 1 + e^{\Theta_1} + e^{\Theta_2} + e^{\Theta_3} + e^{\Theta_4} + a_{12}e^{\Theta_1+\Theta_2} + a_{13}e^{\Theta_1+\Theta_3} + \quad (6.20)$$

$$a_{14}e^{\Theta_1+\Theta_4} + a_{23}e^{\Theta_2+\Theta_3} + a_{24}e^{\Theta_2+\Theta_4} + a_{34}e^{\Theta_3+\Theta_4} +$$

$$b_{123}e^{\Theta_1+\Theta_2+\Theta_3} + b_{124}e^{\Theta_1+\Theta_2+\theta_4} + b_{134}e^{\Theta_1+\Theta_3+\Theta_4} +$$

$$b_{234}e^{\Theta_2+\Theta_3+\Theta_4} + c_{1234}e^{\Theta_1+\Theta_2+\Theta_3+\Theta_4}.$$

The coefficients a_{ij}, $i < j$ are given by Eq. (6.18), while $b_{ijr} = a_{ij}a_{ir}a_{jr}$ for $1 \leq i < j < r \leq 4$, $c_{1234} = a_{12}a_{13}a_{14}a_{23}a_{24}a_{34}$.

We shall say several words about the one and two soliton solutions of the nonlinear Kaup-Kupershmidt equation [77], [84], [123] (see also [89], [124]):

$$u_t + 10uu_{xxx} + 25u_xu_{xx} + 20u^2u_x + u_{xxxx} = 0. \quad (6.21)$$

Equation (6.21) possesses bilinear representation but the explicit computations of the N soliton solutions are technically very difficult and tiresome. The simplified version of Hirota's method was applied to Eq. (6.21) in [60] and for the first time $N = 1, 2, 3$ soliton solutions were found there. We follow closely [60] for the cases $N = 1, 2$.

We look for a solution of Eq. (6.21) having the form

$$u = \frac{3}{2}(ln\ f)_{xx}. \quad (6.22)$$

(See the difference with Eq. (6.2), where 2 stands as a factor.)

Then instead of Eq. (6.4), Eq. (6.14) we come to the following sixth order nonlinear PDE:

$$4f^3(f_{xt} + f_{xxxxxx}) - f^2(4f_tf_x - 5f_{xxx}^2 + 24f_xf_{xxxxx}) - \quad (6.23)$$

$$30f(f_x f_{xx} f_{xxx} - 2f_x^2 f_{xxxx}) + 15(3f_x^2 f_{xx}^2 - 4f_x^3 f_{xxx}) = 0.$$

Define now the linear differential operator $\mathcal{L}$:

$$\mathcal{L} = 4\left(\frac{\partial^2}{\partial x \partial t} + \frac{\partial^6}{\partial x^6}\right) \tag{6.24}$$

and the nonlinear (bilinear) operators

$$N_1(f,g) = -4f_t g_x + 5f_{xxx} g_{xxx} - 24f_x g_{xxxxx} \tag{6.25}$$

$$N_2(f,g.h) = -30f_x g_{xx} h_{xxx} + 60f_x g_x h_{xxxx}$$

$$N_3(f,g,h,j) = 45f_x g_x h_{xx} j_{xx} - 60f_x g_x h_x j_{xxx}.$$

Therefore, Eq. (6.23) can be rewritten as

$$f^3 \mathcal{L}(f) + f^2 N_1(f,f) + f N_2(f,f,f) + N_3(f,f,f,f) = 0. \tag{6.26}$$

Equation (6.26) is a third order polynomial of f with coefficients nonlinear PDE in (t,x) of the derivatives of f.

Seeking a solution of Eq. (6.26) of the type Eq. (6.6) and equating the coefficients of different powers ε^j, $j \geq 1$ to 0 we get:

(ã) $O(\varepsilon) : \mathcal{L}f^{(1)} = 0$ $\hspace{4cm}$ (6.27)

(b̃) $O(\varepsilon^2) : \mathcal{L}f^{(2)} = -N_1(f^{(1)}, f^{(1)})$

(c̃) $O(\varepsilon^3) : \mathcal{L}f^{(3)} = -3f^{(1)}\mathcal{L}f^{(2)} - 2f^{(1)}N_1(f^{(1)}, f^{(1)}) - N_1(f^{(2)}, f^{(1)})$
$-N_1(f^{(1)}, f^{(2)}) - N_2(f^{(1)}, f^{(1)}, f^{(1)})$

(d̃) $O(\varepsilon^4) : \mathcal{L}f^{(4)} = -3f^{(1)}\mathcal{L}f^{(3)} - 3f^{(2)}\mathcal{L}f^{(2)} - 3(f^{(1)})^2\mathcal{L}f^{(2)}$
$-N_1(f^{(1)}, f^{(3)}) - N_1(f^{(3)}, f^{(1)}) - N_1(f^{(2)}, f^{(2)}) - 2f^{(1)}N_1(f^{(1)}, f^{(2)})$
$-2f^{(1)}N_1(f^{(2)}, f^{(1)}) - 2f^{(2)}N_1(f^{(1)}, f^{(1)}) - (f^{(1)})^2 N_1(f^{(1)}, f^{(1)})$
$-N_2(f^{(1)}, f^{(2)}, f^{(1)}) - N_2(f^{(2)}, f^{(1)}, f^{(1)}) - f^{(1)}N_2(f^{(1)}, f^{(1)}, f^{(1)})$
$-N_3(f^{(1)}, f^{(1)}, f^{(1)}, f^{(1)}).$

The last expression seems very heavy and contains 15 summands.

Consider the case $N = 1$, i.e. $f^{(1)} = e^\Theta$, $\Theta = kx - wt + \delta$.

Thus, $\mathcal{L}f^{(1)} = 0 \Rightarrow \mathcal{L}(k, -w) = 0 \iff 4(-kw + k^6) = 0 \iff w = k^5$, $k \neq 0$.

To solve Eq. (6.27) case (b̃) we compute

$$-N_1(e^\Theta, e^\Theta) = 15k^6 e^{2\Theta}.$$

Taking $f^{(2)} = ae^{2\Theta}$ we have

$$\mathcal{L}f^{(2)} = a240k^6 e^{2\Theta} = 15k^6 e^{2\Theta},$$

i.e. $a = \frac{1}{16} \Rightarrow f^{(2)} = \frac{1}{16}e^{2\Theta}.$

One can check that $f^{n)} = 0$ for $n \geq 3$ and consequently one soliton solution of Eq. (6.21) is the following:

$$u = \frac{3}{2}(\ln f_2)_{xx}, f_2 = 1 + e^{\Theta} + \frac{1}{16}e^{2\Theta}, \Theta = kx - k^5 t + \delta.$$

Here we have taken $\varepsilon = 1$. Thus,

$$u = 24k^2 \frac{e^{\Theta}(4e^{\Theta} + e^{2\Theta} + 16)}{(16e^{\Theta} + e^{2\Theta} + 16)^2}.$$

The situation for $N = 2$ is more complicated. We give here a short sketch of the computations.

1). Let $f^{(1)} = e^{\Theta_1} + e^{\Theta_2}$, $\Theta_i = k_i x - k_i^5 t + \delta_i$, $k_i \neq 0$, $i = 1, 2$, $k_1 \neq k_2$. Then $-N_1(f^{(1)}, f^{(1)}) = 15k_1^6 e^{2\Theta_1} + 15k_2^6 e^{2\Theta_2} + 10k_1 k_2(2k_1^4 - k_1^2 k_2^2 + 2k_2^4)e^{\Theta_1 + \Theta_2}$ and consequently $f^{(2)}$ must be written as $f^{(2)} = ae^{2\Theta_1} + be^{2\Theta_2} + a_{12}e^{\Theta_1 + \Theta_2}$. From Eq. (6.27) case $(\tilde{b})$ we have: $\mathcal{L}f^{(2)} = -N_1(f^{(1)}, f^{(1)})$ and $\mathcal{L}f^{(2)} = 240ak_1^6 e^{2\Theta_1} + 240bk_2^6 e^{2\Theta_2} + 20a_{12}k_1 k_2(k_1 + k_2)^2 (k_1^2 + k_1 k_2 + k_2^2)e^{\Theta_1 + \Theta_2}$. Therefore, $240ak_1^6 = 15k_1^6$, $240bk_2^6 = 15k_2^6$, $20a_{12}k_1 k_2(k_1 + k_2)^2 \times (k_1^2 + k_1 k_2 + k_2^2) = 10k_1 k_2(2k_1^4 - k_1^2 k_2^2 + 2k_2^4)$, i.e.

$$f^{(2)} = \frac{1}{16}e^{2\Theta_1} + \frac{1}{16}e^{2\Theta_2} + \frac{1}{2}\frac{2k_1^4 - k_1^2 k_2^2 + 2k_2^4}{(k_1 + k_2)^2(k_1^2 + k_1 k_2 + k_2^2)}e^{\Theta_1 + \Theta_2}.$$

2). Computing the left-hand side of Eq. (6.27) case $(\tilde{c})$ we see that $f^{(3)}$ should exist in the form $f^{(3)} = b_{12}(e^{\Theta_1 + 2\Theta_2} + e^{2\Theta_1 + \Theta_2})$.

After some elementary algebra we find that

$$b_{12} = \frac{(k_1 - k_2)^2(k_1^2 - k_1 k_2 + k_2^2)}{16(k_1 + k_2)^2(k_1^2 + k_1 k_2 + k_2^2)}.$$

The most complicated part in that study is the solvability of Eq. (6.27) $(\tilde{d})$. Long computations (in the right-hand side of Eq. (6.27) case $(\tilde{d})$ participate 15 terms containing in a nonlinear way $f^{(1)}$, $f^{(2)}$, $f^{(3)}$) show that $f^{(4)} = c_{12}e^{2\Theta_1 + 2\Theta_2}$, where $c_{12} = b_{12}^2$. Other long computations verify that $f^{(n)} = 0$ for $n \geq 5$. Taking $\varepsilon = 1$ one obtains that the $N = 2$ soliton solution of Eq. (6.21) exists and is given explicitly by the formula

$$f_4 = 1 + e^{\Theta_1} + e^{\Theta_2} + \frac{1}{16}e^{2\Theta_1} + \frac{1}{16}e^{2\Theta_2} + a_{12}e^{\Theta_1 + \Theta_2} +$$

$$b_{12}(e^{2\Theta_1 + \Theta_2} + e^{\Theta_1 + 2\Theta_2}) + b_{12}^2 e^{2\Theta_1 + 2\Theta_2};$$

$$u = \frac{3}{2}(\ln f_4)_{xx}.$$

6.2   Short description of direct Hirota's approach for finding soliton solutions of some classes of nonlinear PDEs

R. Hirota developed in the 70's of the last century the so-called direct methods that enable the finding of explicit solutions (solitons) of some evolution classes of PDEs. The method in relatively simple, nice and effective in many situations.

The direct method is based on two main ideas:

1). To find an appropriate change of the function satisfying the corresponding nonlinear PDE and to transform this evolution equation into the Hirota's bilinear form. The form is quadratic with respect to the function (its derivatives).

2). To find formal series from the perturbation theory that satisfy the bilinear equation. In the special case of soliton solutions these series are reduced to finite sums.

In some cases one can use mathematical induction in order to prove that some formula found heuristically in fact satisfies the equation.

Points 1), 2) are usually nontrivial. As it concerns our investigations here, we shall restrict ourselves to the logarithmic change $u = 2(ln \ f)_{xx}$ only.

Following Hirota and other authors discussing and applying his approach: [3], [5], [42], [61], [62], [63], [64], [65], [102] we introduce the following new binary differential operator D acting on pair of functions a, b:

$$D_t^n D_x^m a.b = \left(\frac{\partial}{\partial t} - \frac{\partial}{\partial t'}\right)^n \left(\frac{\partial}{\partial x} - \frac{\partial}{\partial x'}\right)^m a(x,t)b(x',t')|_{t=t',x=x'}, m, n \in \mathbf{N_0}.$$

$$(6.28)$$

If δ, ε are two constants then equality $\frac{\partial}{\partial z} = \delta\frac{\partial}{\partial t} + \varepsilon\frac{\partial}{\partial x}$ enables us to define $D_z = \delta D_t + \varepsilon D_x$. Equation (6.28) is given in the case of 2 independent variables only but it can be generalized for several variables too. From the Definition Eq. (6.28) it follows immediately that $D_x G \cdot F = G_x F - F_x G$, $D_t G \cdot F = G_t F - GF_t$, $D_x^2 F \cdot F = 2(F_{xx}F - F_x^2)$. As usual D_x, D_t are Hirota's derivatives, while $G_x = \frac{\partial G}{\partial x}$, $F_t = \frac{\partial F}{\partial t}$ etc.

We give below a list of properties of the operator D.

1. $D_t^m D_x^n a.b = D_x^n D_t^m a.b = D_x^{n-1} D_t^m D_x a.b$
2. $D_t^m D_x^n a.1 = \partial_t^m \partial_x^n a$
3. $(D_t + \varepsilon D_x)^n a.b = D_t^n a.b + n\varepsilon D_t^{n-1} D_x a.b + \ldots + \varepsilon^n D_x^n a.b$
   (binomial formula)
4. $D_z^m a.b = (-1)^m D_z^m b.a$

5. $D_z^m a.a = 0$ for m-odd
6. $D_z^m a.a = 2D_z^{m-1} a_z.a$ for m-even, $a_z = \frac{\partial a}{\partial z}$.
7. $D_x D_t a.a = 2D_x a_t.a = 2D_t a_x.a$.
8. $D_x^m e^{p_1 x}.e^{p_2 x} = (p_1 - p_2)^m e^{(p_1+p_2)x} \Rightarrow D_x^m e^{p_1 x}.e^{p_1 x} = 0$.

Assume that $F(D_t, D_x)$ is a polynomial in D_t, D_x (Hirota's operator) $(F(w,p))$. Then

9.
$$F(D_t, D_x)e^{p_1 x+w_1 t}.e^{p_2 x+w_2 t} = \frac{F(w_1-w_2,p_1-p_2)}{F(w_1+w_2,p_1+p_2)} F(\partial_t, \partial_x)e^{(w_1+w_2)t+(p_1+p_2)x}$$

10. $\frac{\partial^2}{\partial z^2}\ln f = \frac{D_z^2 f.f}{2f^2}$

11. $\frac{\partial^4}{\partial z^4}\ln f = \frac{D_z^4 f.f}{2f^2} - 6[\frac{D_z^2 f.f}{2f^2}]^2$
Corollary from 11.:

12. $2\frac{\partial^2}{\partial x^2}\ln f = \frac{D_x^2 f.f}{f^2}$, $2\frac{\partial^2}{\partial x \partial t}\ln f = \frac{D_x D_t f.f}{f^2}$, $2\frac{\partial^4}{\partial x^4}\ln f = \frac{D_x^4 f.f}{f^2} - 3(\frac{D_x^2 f.f}{f^2})^2$.

We shall prove 8. and 10. only because of the lack of space.
According to Eq. (6.28)

$$D_x^m e^{p_1 x}.e^{p_2 x} = \left(\frac{\partial}{\partial x} - \frac{\partial}{\partial x'}\right)^m e^{p_1 x+p_2 x'}\Big|_{x=x'} =$$

$$\left(\frac{\partial^m}{\partial x^m} - m\frac{\partial^{m-1}}{\partial x^{m-1}}\frac{\partial}{\partial x'} + \ldots + (-1)^m\frac{\partial^m}{\partial x'^m}\right)e^{p_1 x+p_2 x'}\Big|_{x=x'} =$$

$$(p_1^m - mp_1^{m-1}p_2 + \ldots + (-1)^m p_2^m)e^{(p_1+p_2)x} = (p_1-p_2)^m e^{(p_1+p_2)x} =$$

$$\frac{(p_1-p_2)^m}{(p_1+p_2)^m}\partial_x^m(e^{(p_1+p_2)x}).$$

As it concerns 10.

$$D_x^2 f.f = \left(\frac{\partial}{\partial x} - \frac{\partial}{\partial x'}\right)^2 f(x)f(x')|_{x=x'} =$$

$$\left(\frac{\partial^2}{\partial x^2} - 2\frac{\partial^2}{\partial x \partial x'} + \frac{\partial^2}{\partial x'^2}\right)f(x)f(x')|_{x=x'} =$$

$$2\left(\frac{\partial^2}{\partial x^2}f(x)\right)f(x) - 2\left(\frac{\partial f}{\partial x}\right)^2 = 2(f''f - (f')^2).$$

On the other hand,

$$\frac{\partial^2}{\partial x^2}\ln f = \left(\frac{f'}{f}\right)_x = \frac{f''f - (f')^2}{f^2} \Rightarrow \frac{\partial^2}{\partial x^2}\ln f = \frac{D_x^2 f.f}{2f^2}.$$

The identity 11. can be proved in a similar way up to some tiresome computations.

The logarithmic transformation enables us to rewrite Eq. (6.1) into a bilinear form. Starting from Eq. (6.3) with $w = 2(ln f)_x \Rightarrow u = 2(ln f)_{xx}$ we have via 10-11

$$w_t = 2(ln f)_{xt} = \frac{D_x D_t f \cdot f}{f^2}, w_x^2 = 4[(ln f)_{xx}]^2 = \left(\frac{D_x^2 f \cdot f}{f^2}\right)^2,$$

$$w_{xxx} = 2(ln f)_{xxxx} = \frac{D_x^4 f \cdot f}{f^2} - 3\left(\frac{D_x^2 f \cdot f}{f^2}\right)^2,$$

i.e. Eq. (6.3) takes the form:

$$(D_{xt}^2 + D_x^4)f \cdot f = 0. \tag{6.29}$$

Consider now the Boussinesq equation Eq. (6.13). We shall show below that Eq. (6.13) can be written in the bilinear form:

$$(D_t^2 - D_x^2 - D_x^4)f \cdot f = 0, u = 2(ln f)_{xx}. \tag{6.30}$$

In fact, as we know $u = \frac{D_x^2 f \cdot f}{f^2}$, $2\frac{\partial^2}{\partial t^2} ln f = \frac{D_t^2 f \cdot f}{f^2}$, $u_{xx} = 2(ln f)_{xxxx}$ are given by 12. Substituting these expressions in Eq. (6.30) we get:

$$2f^2(ln f)_{tt} - f^2 u - f^2 u_{xx} - f^2 3u^2 = 0, \tag{6.31}$$

i.e.

$$0 = 2(ln f)_{tt} - u - u_{xx} - 3u^2 \Rightarrow u_{tt} - u_{xx} - u_{xxxx} - 3(u^2)_{xx} = 0. \tag{6.32}$$

This way we obtain the nonlinear PDE Eq. (6.13). Conversely, Eq. (6.13) leads to Eq. (6.30).

Our further step is to consider the 2D Kadomtsev-Petviashvili equation [73] (see also [5], [102]):

$$(u_t + 6uu_x + u_{xxx})_x + u_{yy} = 0. \tag{6.33}$$

Equation (6.33) can be transformed in the Hirota's bilinear equation

$$(D_t D_x + D_y^2 + D_x^4 f) \cdot f = 0 \tag{6.34}$$

after the logarithmic change $u = 2(ln f)_{xx}$.

To do this we substitute the expressions for $D_{tx}^2 f \cdot f$, $D_x^4 f \cdot f$, $D_y^2 f \cdot f$ from 12 into Eq. (6.34) and differentiate two times in x the corresponding PDE

$$2(ln f)_{xt} + 2(ln f)_{xxxx} + 3u^2 + 2(ln f)_{yy} = 0.$$

Therefore, we come to Eq. (6.33). Certainly, from Eq. (6.33) we obtain Eq. (6.34) too.

The last equation to be studied is the Vakhnenko equation [131]

$$\frac{\partial}{\partial x}\left(\frac{\partial}{\partial t} + u\frac{\partial}{\partial x}\right)u + u = 0, \tag{6.35}$$

which governs the propagation of waves in a relaxing medium.

We introduce the new independent variables

$$\begin{vmatrix} x = \Theta(X,T) = T + \int_{-\infty}^{X} U(X',T)dX', \\ t = x, \end{vmatrix} \tag{6.36}$$

where $u(x,t) = U(X,T)$.

Then $\frac{\partial U}{\partial X} = \frac{\partial u}{\partial t} + \frac{\partial u}{\partial x}\frac{\partial x}{\partial X} = \frac{\partial u}{\partial t} + u\frac{\partial u}{\partial x}$, $\frac{\partial U}{\partial T} = \Phi(X,T)\frac{\partial u}{\partial x}$, $\Phi = 1 + \int_{-\infty}^{X} \frac{\partial U}{\partial T}dX'$, $\Phi(-\infty,T) = 1$. Thus $\Phi_X = U_T$.

The Eq. (6.35) takes the form:

$$U_{XT} + \Phi U = 0, \tag{6.37}$$

as $\frac{\partial}{\partial x} = \frac{1}{\Phi}\frac{\partial}{\partial T}$, $\frac{\partial}{\partial t} + u\frac{\partial}{\partial x} = \frac{\partial}{\partial X} \Rightarrow \Phi = -\frac{U_{XT}}{U}$.

Differentiating Eq. (6.37) with respect to X we get:

$$UU_{TXX} - U_X U_{XT} + U^2 U_T = 0, \tag{6.38}$$

as $\Phi_X = U_T$.

Put $W_X = U$ assuming that all the corresponding integrals are convergent. Then $\Phi_X = U_T = W_{XT} \Rightarrow \Phi = W_T + A(T)$, $\Phi(-\infty,T) = A(T) = 1 \Rightarrow \Phi = W_T + 1$. Then Eq. (6.37) becomes

$$W_{XXT} + W_X W_T + W_X = 0. \tag{6.39}$$

The equations Eq. (6.38), Eq. (6.39) are equivalent. Therefore,

$$W = 6(lnf)_X \Rightarrow U = W_X = 6(lnf)_{XX} = \frac{3D_X^2 f.f}{f^2}, \tag{6.40}$$

$W_T = 6(lnf)_{XT} = \frac{3D_{XT}^2 f.f}{f^2}$, $W_{XXT} = 6(lnf)_{XXXT} = 3\frac{D_T D_X^3 f.f}{f^2}$ $-9\frac{(D_X^2 f.f)(D_{XT}^2 f.f)}{f^4}$ (see 12).

Thus Eq. (6.39) takes the bilinear form

$$F(D_x, D_t)f.f = (D_T D_X^3 + D_X^2)f.f = 0. \tag{6.41}$$

We must solve Eq. (6.41) by using Hirota's method and then we shall obtain $u(x,t)$ into a parametric form:

$$\begin{vmatrix} u(x,t) = U(t,T) = U(X,T) \\ x = \Theta(t,T) \end{vmatrix}$$

where $\Theta = T + W(X,T) + x_0$.

Our next step is to solve the bilinear Hirota's equations Eq. (6.29), Eq. (6.30), Eq. (6.34) via the perturbation method. Fortunately, in these three cases we shall obtain solutions written as finite sums of linear exponents.

6.3　Bilinear equations of the type $\mathcal{L}(D_x, D_t)f \cdot f = 0$

In this section we shall assume that $\mathcal{L}$ is an even order polynomial $\mathcal{L}(k, w)$. We shall solve the bilinear equations via the perturbation method i.e. we are trying to find f of the form Eq. (6.6). Substituting Eq. (6.6) into the bilinear Eq. (6.29) and equating to zero the coefficients in front of the powers of ε we get:

$$
\begin{aligned}
\varepsilon^1: \quad & 2\tfrac{\partial}{\partial x}\left(\tfrac{\partial}{\partial t} + \tfrac{\partial^3}{\partial x^3}\right)f^{(1)} = 2\mathcal{L}(\partial_x, \partial_t)f^{(1)} = 0, \mathcal{L}(D_x, D_t) = D_x D_t + D_x^4, \\
\varepsilon^2: \quad & 2\tfrac{\partial}{\partial x}\left(\tfrac{\partial}{\partial t} + \tfrac{\partial^3}{\partial x^3}\right)f^{(2)} = -\mathcal{L}(D_x, D_t)f^{(1)} \cdot f^{(1)} \\
\varepsilon^3: \quad & 2\tfrac{\partial}{\partial x}\left(\tfrac{\partial}{\partial t} + \tfrac{\partial^3}{\partial x^3}\right)f^{(3)} = \mathcal{L}(D_x, D_t)(f^{(3)} \cdot 1 + 1 \cdot f^{(3)}) = \\
& -\mathcal{L}(D_x, D_t)(f^{(1)} \cdot f^{(2)} + f^{(2)} \cdot f^{(1)}) = -2\mathcal{L}(D_x, D_t)f^{(1)} \cdot f^{(2)} \qquad (6.42) \\
\varepsilon^4: \quad & 2\tfrac{\partial}{\partial x}\left(\tfrac{\partial}{\partial t} + \tfrac{\partial^3}{\partial x^3}\right)f^{(4)} = -\mathcal{L}(D_x, D_t)(f^{(3)} \cdot f^{(1)} + f^{(2)} \cdot f^{(2)} + f^{(1)} \cdot f^{(3)})
\end{aligned}
$$

....

We used the above given properties 4. and 2. of the Operator D.

We take $f^{(1)} = e^{\eta_1}$, $\eta_1 = k_1 x + w_1 t$ we solve $\mathcal{L}(\partial_x, \partial_t)f^{(1)} = 0$ as for $k_1^3 = -w_1$, $\mathcal{L}(k_1, w_1) = 0$.

Then $f^{(1)} = e^{\eta_1} + e^{\eta_2}$, $\eta_i = k_i x + w_i t$, $w_i + k_i^3 = 0$ is a solution of the same PDE.

In order to find a two soliton solution we apply property 8. of D computing the expression:

$$\mathcal{L}(D_x, D_t)f^{(1)} \cdot f^{(1)} = (D_t D_x + D_x^4)f^{(1)} \cdot f^{(1)} =$$

$$(D_t D_x + D_x^4)(e^{\eta_1} + e^{\eta_2}) \cdot (e^{\eta_1} + e^{\eta_2}) =$$

$$2(D_x D_t + D_x^4)e^{\eta_1} \cdot e^{\eta_2} = 2\mathcal{L}(k_1 - k_2, w_1 - w_2)e^{\eta_1 + \eta_2} =$$

$$2(k_1 - k_2)(w_1 - w_2 + (k_1 - k_2)^3)e^{\eta_1 + \eta_2}.$$

Taking $f^{(2)} = a_{12}e^{\eta_1 + \eta_2}$ in the second equation of Eq. (6.42) we conclude that

$$a_{12} = \frac{2\mathcal{L}(k_1 - k_2, w_1 - w_2)}{2\mathcal{L}(k_1 + k_2, w_1 + w_2)} = \frac{(k_1 - k_2)^2}{(k_1 + k_2)^2}.$$

We shall show that $f^{(1)} + f^{(2)}$ satisfies the third equation of Eq. (6.42). In fact, $\frac{1}{a_{12}}\mathcal{L}(D_x, D_t)f^{(1)} \cdot f^{(2)} = \mathcal{L}(D_x, D_t)e^{\eta_1 + \eta_2} \cdot (e^{\eta_1} + e^{\eta_2})$ $= \mathcal{L}(k_2, w_2)e^{2\eta_1 + \eta_2} + \mathcal{L}(k_1, w_1)e^{\eta_1 + 2\eta_2} = 0$.

Consequently we can choose $f^{(3)} = 0$. One can easily see that $\mathcal{L}(D_x, D_t)f^{(2)} \cdot f^{(2)} = 0$. This way we conclude that $f_2 = 1 + \varepsilon f^{(1)} + \varepsilon^2 f^{(2)}$ is a 2 soliton solution of Eq. (6.29). f_2 obtained by Hirota's method coincides with f_2 in Eq. (6.7)(b).

We propose below a slight modification and generalization of that result in the multidimensional case.

Thus, let the variables be $(x,t) = (x_1, \ldots, x_n, t)$ and the corresponding bilinear equation is

$$F(D_x, D_t)f.f = 0, D_x = (D_{x_1}, \ldots, D_{x_n}). \tag{6.43}$$

F is a polynomial with respect to (k, w), $k = (k_1, \ldots, k_n)$, $w \in \mathbf{R}^1$.

Proposition 6.1.

Consider the equation Eq. (6.43) and assume that

(i) $F(-D_x, -D_t) = F(D_x, D_t)$

(ii) $F(0,0) = 0$.

Then Eq. (6.43) possesses the two soliton solution $f = 1 + e^{\eta_1} + e^{\eta_2} + a_{12}e^{\eta_1 + \eta_2}$, *where* $\eta_i = <k_i, x> + w_i t$, $i = 1, 2$, $k_i \in \mathbf{R}^n$, $w_i \in \mathbf{R}$ *satisfy the dispersion relation* $F(k_i, w_i) = 0$, $i = 1, 2$ *and* $a_{12} = -\frac{F(k_1 - k_2, w_1 - w_2)}{F(k_1 + k_2, w_1 + w_2)}$.

Due to 9. the proof is trivial. Condition (i) is not essential as each polynomial $F(D) = F_{odd}(D) + F_{even}(D)$ and $F_{odd}(D)f.f = 0$ according to 5. If (ii) is violated the ε^0 term in Eq. (6.42) is $F(0,0)1 \neq 0$.

The N soliton solution f_N of Eq. (6.29) exists in the following form. Put $a_{ij} = e^{A_{ij}}$. Then f_N may be written as

$$f_N = \sum_{\mu=0,1} exp\left[\sum_{i=1}^{N} \mu_i \eta_i + \sum_{i<j}^{N} A_{ij}\mu_i\mu_j\right]. \tag{6.44}$$

Here $\sum_{\mu=0,1}$ stands for the summation over all possible combinations of $\mu_1 = 0, 1$; $\mu_2 = 0, 1$; $\ldots \mu_N = 0, 1$. $\sum_{i<j}^{N}$ means the summation is over the pairs $(i,j) \in \{1, 2, \ldots, N\}$ such that $1 \leq i < j \leq N$. Hence, for $N = 3$, $f_3 = 1 + e^{\eta_1} + e^{\eta_2} + e^{\eta_3} + e^{A_{12} + \eta_1 + \eta_2} + e^{A_{13} + \eta_1 + \eta_3} + e^{A_{23} + \eta_2 + \eta_3} + e^{A_{12} + A_{13} + A_{23} + \eta_1 + \eta_2 + \eta_3}$.

This formula was given by R. Hirota in 1971.

An inductive proof of the above mentioned result is proposed in [5]. The proof is complicated and tiresome.

Theorem 6.1. *(Hirota)*

Consider the multidimensional bilinear equation Eq. (6.43) under the conditions (i), (ii) and the dispersive relations $F(k_i, w_i) = 0$, $i = 1, 2, \ldots N$. *Put* $e^{A_{ij}} = -\frac{F(k_i - k_j, w_i - w_j)}{F(k_i + k_j, w_i + w_j)}$.

Assume that the following Hirota's condition ([63]) holds:

(iii) $\sum_{\sigma=0,1} F(\sum_{i=1}^{N} \sigma_i k_i, \sum_{i=1}^{N} \sigma_i w_i) \prod_{1 \leq i < j}^{N} F(\sigma_i k_i - \sigma_j k_j, \sigma_i w_i - \sigma_j w_j)\sigma_i\sigma_j = 0$,

$k_i \in \mathbf{R}^n$, $w_i \in \mathbf{R}$ *and the summation* $\sum_{\sigma=0,1}$ *is over all possible combinations of* $\sigma_1 = 0, 1$; $\sigma_2 = 0, 1$; $\ldots \sigma_N = 0, 1$.

Then Eq. (6.43) possesses an N-soliton solution of the form Eq. (6.44), $\eta_i = < k_i, x > + w_i t$.

In the special case $N = 2$ (iii) holds without additional restrictions imposed on the even polynomial F.

As we know the Kadomtsev-Petviashvili equation Eq. (6.33) is transformed to the bilinear Eq. (6.34). As we mentioned in 3.3 Satsuma proved in [122] that Eq. (6.34) has an N-soliton solution f_N that has the form

$$f_N = \sum_{\mu=0,1} exp \left[\sum_{1 \leq i < j}^{N} \mu_i \mu_j A_{ij} + \sum_{i=1}^{N} \mu_i \eta_i \right], \qquad (6.45)$$

where $\eta_i = k_i(x + p_i y - C_i t)$, $C_i = k_i^2 + p_i^2$ and $e^{A_{ij}} = \frac{3(k_i - k_j)^2 - (p_i - p_j)^2}{3(k_i + k_j)^2 - (p_i - p_j)^2}$. f_1, f_2 can be found immediately repeating the procedure used for Eq. (6.29).

As it concerns the Vakhnenko equation the situation is more complicated. There are no difficulties to find 1 and 2 and N soliton solutions of the bilinear Eq. (6.41). In the case $N = 1$ the solution is $f_1 = 1 + e^{2\eta}$, where $\eta = kX - wT + \alpha$. From the dispersion relation $F(2k, -2w) = 0$ one gets $w = \frac{1}{4k} \Rightarrow \eta = k(X - cT) + \alpha$, $c = \frac{1}{4k^2}$ (velocity). Thus, $U(X, T) = 6k^2 sech^2 \eta$. The difficulty is due to the fact that we are interested in the solution $u(x, t)$ of Eq. (6.35) in the "old" coordinates (x, t) and not in (X, T). As we know the 2-soliton solution f_2 in the new coordinates (X, T) is given by

$$f_2 = 1 + e^{2\eta_1} + e^{2\eta_2} + b^2 e^{2(\eta_1 + \eta_2)}, \qquad (6.46)$$

where $\eta_i = k_i X - w_i T + \alpha_i$, $i = 1, 2$ and $b^2 = -\frac{F(2(k_1 - k_2), -2(w_1 - w_2))}{F(2(k_1 + k_2), -2(w_1 + w_2))}$. The dispersion relation $F(2k_i, -2w_i) = 0$ leads to $w_i = \frac{1}{4k_i}$ with $c_i = \frac{1}{4k_i^2}$, $\eta_i = k_i(X - c_i T) + \alpha$. Taking $k_2 > k_1 > 0$ we have $b = \frac{k_2 - k_1}{k_1 + k_2} \sqrt{\frac{k_1^2 + k_2^2 - k_1 k_2}{k_1^2 + k_2^2 + k_1 k_2}} \Rightarrow 0 < b < 1$. Thus, $W = 6(ln f_2)_X \Rightarrow U = W_X = 6(ln f_2)_{XX}$. Again the difficulties in investigating Eq. (6.35) come from the transformation $(X, T) \to (x, t)$. In Chapter 2 we obtain a loop solution of the Vakhnenko equation. So we can expect here that for $N = 1$ the solution $u(x, t)$ is of loop type too. Therefore, for $N = 2$ we shall have to deal with two loop soliton solutions of Eq. (6.35). It concerns the interaction of the loops during their propagation. In [97] N soliton solutions of Eq. (6.41) are found. Their form is the same as Eq. (6.44). Going back to the initial coordinates (x, t) one must study the propagation and interaction of N-loop solutions of the Vakhnenko nonlinear evolution PDE.

6.4 Interaction of 3 waves to Kadomtsev-Petviashvili equation and of two loop solutions of the Vakhnenko equation

Consider the 3-soliton solution of Eq. (6.34). As we know

$$u = 2(ln f_3)_{xx}, \quad f_3 = 1 + e^{\eta_1} + e^{\eta_2} + e^{\eta_3} + e^{A_{12}+\eta_1+\eta_2} +$$

$$e^{A_{13}+\eta_1+\eta_3} + e^{A_{23}+\eta_2+\eta_3} + e^{A_{12}+A_{13}+A_{23}+\eta_1+\eta_2+\eta_3}$$

satisfies Eq. (6.33). We impose the additional conditions $e^{A_{12}} = 0$, $e^{A_{23}} = 0$ in the special case $\sqrt{3}(k_1 - k_2) = p_1 - p_2 > 0$, $\sqrt{3}(k_2 - k_3) = p_2 - p_3 > 0$ $\Rightarrow \sqrt{3}(k_1 - k_3) = p_1 - p_3 \Rightarrow e^{A_{13}} = 0$.

Moreover, $\eta_i = k_i(x + p_i y - C_i t)$, $C_i = k_i^2 + p_i^2$.

Assume that $\begin{vmatrix} k_1 > k_2 > k_3 > 0 \Rightarrow C_1 > C_2 > C_3 > 0 \\ p_1 > p_2 > p_3 > 0. \end{vmatrix}$

One can easily see that

$$u =$$

$$2\frac{e^{\eta_1} + e^{\eta_2} + e^{\eta_3} + (k_1 - k_2)^2 e^{\eta_1+\eta_2} + (k_1 - k_3)^2 e^{\eta_1+\eta_3} + (k_2 - k_3)^2 e^{\eta_2+\eta_3}}{(1 + e^{\eta_1} + e^{\eta_2} + e^{\eta_3})^2}.$$

This is a nice explicit form of the solution.

Exercise 1. Consider the function

$$g(y) = \frac{e^{ay} + e^{by} + Ae^{2ay} + Be^{(a+b)y}}{(1 + e^{ay} + e^{by})^2},$$

where the real constants $ab \neq 0$, $A > 0$, $B > 0$.

Prove that:

I: $a > 0$, $\begin{vmatrix} a \geq b > 0 \Rightarrow lim_{y\to\infty}g(y) = const > 0 \\ a > 0 > b, \end{vmatrix}$, $lim_{y\to-\infty}g(y) = 0$;

II: $a > 0$, $0 < a < b \Rightarrow lim_{|y|\to\infty}g(y) = 0$;

III: $a < 0$, $b < a < 0 \Rightarrow lim_{|y|\to\infty}g(y) = 0$;

IV: $0 > b > a \Rightarrow lim_{y\to-\infty}g(y) = const > 0$, $lim_{y\to\infty}g(y) = 0$; $b > 0 > a \Rightarrow g(y) \sim 0$ at $+\infty$, $g(y) \sim const$ at $-\infty$.

Exercise 2. Consider $h(y) = \frac{1+e^{ay}+e^{by}+Ae^{(a+b)y}}{(1+e^{ay}+e^{by})^2}$, $A > 0$ and prove that

V: $lim_{y\to\infty}g(y) = 0$, $lim_{y\to-\infty}g(y) = 1$ for $0 < a < b$;

VI: $lim_{|y|\to\infty}g(y) = 0$ for $ab < 0$ and

VII: $lim_{y\to\infty}g(y) = 0$, $lim_{y\to-\infty}g(y) = 1$ for $a < b < 0$.

Let $t = 0$. Put $l_1 : x = -p_1 y$, $l_2 : x = -p_2 y$, $l_3 : x = -p_3 y$, l_i are straight lines in the plane. Evidently $\eta_1 = \eta_2$ at the straight line $l_4 : x = -\alpha_1 y$, $\alpha_1 = p_1 + k_2\sqrt{3} > 0$ and $\eta_1|_{l_4} = \eta_2|_{l_4} = -k_1 k_2 \sqrt{3} y$. Define $l_5 : x = -\alpha_2 y$,

$\alpha_2 = p_1 + \sqrt{3}k_3$. Then $\eta_1|_{l_5} = \eta_3|_{l_5} = -k_1 k_3 \sqrt{3}y$, while $\eta_2 = \eta_3$ at the line $l_6 : x = -\alpha_3 y$, $\alpha_3 = p_2 + \sqrt{3}k_3$, $\eta_2|_{l_6} = \eta_3|_{l_6} = -k_2 k_3 \sqrt{3}y$. It is easy to see that $lim_{y\to\infty} u|_{l_1} = const > 0$, $lim_{y\to-\infty} u|_{l_1} = const \geq 0$. We have $= 0$ iff $k_2(p_2 - p_1) = k_3(p_3 - p_1)$. Simple calculations show that $lim_{|y|\to\infty} u|_{l_2} = 0$, while $lim_{y\to-\infty} u|_{l_3} = const > 0$, $lim_{y\to+\infty} u|_{l_3} = 0$ (see Ex. 2).

According to the Exercise 1: $lim_{y\to\infty} u|_{l_i} = 0$, $i = 3, 4, 5$, $lim_{y\to-\infty} u|_{l_i} = const > 0$.

The physical interpretation is the following one.

We have propagation of 3 nonlinear waves ("solitons") given by e^{η_1}, e^{η_2}, e^{η_3} and under a triple resonance. Fix $t = -1, -1/2, 0, 1, 1/2$ etc. In the special case $t = 0$ the soliton $u|_{l_2}$, the antikink $u|_{l_3}$ and the kink $u|_{l_1}$ due to their interaction at the origin 0 give rise to 3 new born kinks for $x \to +\infty$. One can analyse the situation for $t \neq 0$, t-fixed too (see [2]).

Concerning the interaction of two loops (specific soliton type solutions of Vakhnenko equation Eq. (6.35)) the picture is rather exotic (for detailed proof one can see [24]).

There are three different cases that appear, namely:

(1) the two loops exchange their amplitudes during the interaction but never overlap

(2) the two loops exchange their amplitudes during the interaction and for a part of the interaction the loops overlap (partial overlapping). The smaller one penetrates into the larger one and then leaves it.

(3) the larger loop catches up the smaller one which during the interaction travels clockwise around the larger loop. After that the smaller loop is ejected behind the larger one (see Figs. 6.2, 6.3).

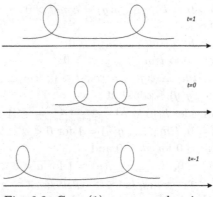

Fig. 6.2. Case (1) — no overlapping.

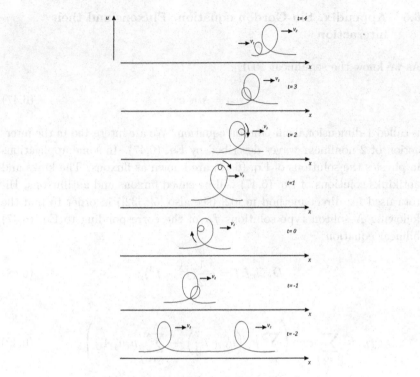

Fig. 6.3. Case (3) — full overlapping.

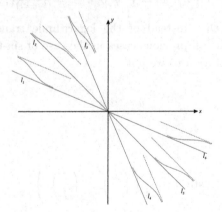

Fig. 6.4. Interaction of kinks for $x \to -\infty$ and new born 3 kinks for $x \to \infty$.

6.5 Appendix. Sin-Gordon equation. Fluxons and their interaction

As we know the semilinear PDE

$$u_{xt} = sin\ u \tag{6.47}$$

is called 1-dimensional sin-Gordon equation. We are interested in the interaction of 2 nonlinear waves described by Eq. (6.47). In some applications in physics the solutions of Eq. (6.47) are known as fluxons. The kinks and antikinks solutions of Eq. (6.47) will be called fluxons and antifluxons. Hirota used his direct method in [62] (see also [5], [32]) in order to find the following N-soliton type solutions f_N of the corresponding to Eq. (6.47) bilinear equation

$$D_x D_t f.f = -\frac{1}{2}(\bar{f}^2 - f^2) : \tag{6.48}$$

$$f_N = \sum_{\mu=0,1} exp\left(\sum_{j=1}^{N} \mu_j\left(\eta_j + i\frac{\pi}{2}\right) + \sum_{1 \le i < j}^{N} \mu_i \mu_j A_{ij}\right), \tag{6.49}$$

where $\eta_j = k_j x - w_j t + \eta_j^0$, $w_j = \frac{1}{k_j} \ne 0$, $e^{A_{ij}} = \frac{(k_i - k_j)^2}{(k_i + k_j)^2}$, $k_i \ne k_j$.

We point out that instead of the logarithmic transformation $u = 2(lnf)_{xx}$ used in the all previous cases in the case of sin-Gordon equation (complex valued solution u) we put

$$u = 2iln\frac{\bar{f}}{f} \tag{6.50}$$

that implies

$$sin\ u = \frac{1}{2i}\left(\left(\frac{f}{\bar{f}}\right)^2 - \left(\frac{\bar{f}}{f}\right)^2\right).$$

We do not enter into technical details and do not prove here that the Eq. (6.47) is reduced via the transformation Eq. (6.50) to Eq. (6.48). The corresponding calculations are given in [5], [62]. Let $f = F + iG$ (F, G-real valued functions).

Then Eq. (6.50) leads to

$$u = 4arg\, f = 4arctg\frac{G}{F},\qquad (6.51)$$

as $lnz = ln|z| + iarg\, z$, $arg\frac{\bar{f}}{f} = -2arg\, f$. Thus, $N = 1 \Rightarrow u_1 = 4arctge^{\eta_1}$, as $f_1 = 1+ie^{\eta_1}$, while $f_2 = 1+i(e^{\eta_1}+e^{\eta_2})+e^{A_{12}+\eta_1+\eta_2+i\pi} = 1-e^{A_{12}+\eta_1+\eta_2}+i(e^{\eta_1}+e^{\eta_2}) \Rightarrow u_2 = 4arctg\frac{e^{\eta_1+\eta_2}}{1-e^{A_{12}+\eta_1+\eta_2}}$ is the 2-soliton type solution of Eq. (6.47).

The strictly monotonically increasing function $u_1(x)$ for $k_1 > 0$, t-fixed is a fluxon as $u_1(x) \to_{x\to-\infty} 0$, $u_1(x) \to_{x\to\infty} 2\pi$.

In a similar way one sees that $u_1(x)$, t-fixed is an antifluxon for $k_1 < 0$.

Consider now the interaction of 2 fluxons with $k_2 > k_1 > 0 \Rightarrow$ the velocities $c_1 = \frac{1}{k_1^2} > c_2 = \frac{1}{k_2^2} > 0$, $\eta_1 = k_1(x - c_1t)$, $\eta_2 = k_2(x - c_2t)$.

On their way they have for finite (t, x) an obstacle and therefore "de facto" a double collision appears.

(A) Fix $\begin{cases} \eta_1 = const. \\ \eta_1^0 = 0 = \eta_2^0 \end{cases}$. Then $\begin{aligned} x \sim c_1t(x \to \infty &\iff t \to \infty) \\ |x| \gg 1(x \to -\infty &\iff t \to -\infty) \end{aligned}$ and $\eta_2 = k_2(x - c_2t) \sim k_2(c_1 - c_2)t$, $c_1 - c_2 > 0$.

Therefore, $t \to -\infty \Rightarrow e^{\eta_2} \to +0 \Rightarrow 4arctge^{\eta_1} > 0$ (I), while $t \to \infty \Rightarrow e^{\eta_2} \to +\infty \Rightarrow u_2 \sim -4arctg(e^{-A_{12}-\eta_1})$ (II).

The identity $arctg\alpha + arctg\frac{1}{\alpha} = sgn\alpha\frac{\pi}{2}$ shows that $u_2 \sim_{t\to\infty} -2\pi + 4arctg(e^{A_{12}+\eta_1})$ (II).

(B) Fix $\eta_2 = const$. Then $x \sim c_2t$, $|x| \gg 1 \Rightarrow \eta_1 \sim k_1(c_2 - c_1)t$, $c_2 - c_1 < 0$ and $t \to -\infty \Rightarrow u_2 \sim -2\pi + 4arctg(e^{A_{12}+\eta_2})$ (III); $t \to +\infty \Rightarrow u_2 \sim 4arctge^{\eta_2}$ (IV).

Geometrically we have (Fig. 6.5):

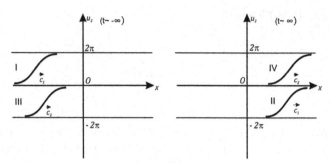

Fig. 6.5.

We are interested now in the interaction of two fluxons for finite time T. So let $1 = e^{\eta_1 + \eta_2 + A_{12}}$, $e^{A_{12}} < 1$ where $e^{A_{12}} \in (0,1)$. Therefore $\eta_1 + \eta_2 = -A_{12} \iff$

$$(k_1 + k_2)x - (w_1 + w_2)t = -A_{12}. \qquad (6.52)$$

Equation (6.52) is the equation of the line L; $L \cap \{|x - c_1 t| \leq \delta \ll 1\}$ is bounded.

Fix some t satisfying Eq. (6.52). Therefore, the corresponding $x(t)$ satisfies

$$x = x(t) = \frac{1}{k_1 k_2}t - \frac{A_{12}}{k_1 + k_2}. \qquad (6.53)$$

Obviously, $x < x(t) \Rightarrow (\eta_1 + \eta_2)(x,t) < (\eta_1 + \eta_2)(x(t),t) = -A_{12} \Rightarrow u_2 > 0$, while $u_2 < 0$ for $x > x(t)$ and

$$\lim_{\substack{x \to x(t) \\ x < x(t)}} u_2 = 2\pi,$$

$$\lim_{\substack{x \to x(t) \\ x > x(t)}} u_2 = -2\pi.$$

Consequently, for every fixed t the function $u_2(x)$ has a jump at $x = x(t)$.

Moreover, $\frac{\partial}{\partial x} \frac{e^{\eta_1} + e^{\eta_2}}{1 - e^{A_{12} + \eta_1 + \eta_2}} = \frac{k_1 e^{\eta_1} + k_2 e^{\eta_2} + e^{A_{12}}(k_2 e^{2\eta_1 + \eta_2} + k_1 e^{2\eta_2 + \eta_1})}{(1 - e^{A_{12} + \eta_1 + \eta_2})^2}$ implies that for fixed t the function $u_2 = u_2(x)$ is strictly monotonically increasing for $x \neq x(t)$, $\lim_{x \to x(t)} \frac{\partial u_2}{\partial x}$ exists and is positive, $\lim_{|x| \to \infty} u_2(x) = 0$.

Below on Fig. 6.6 we propose the corresponding geometrical interpretation of the interaction of two fluxons.

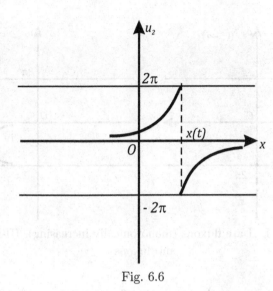

Fig. 6.6

Our next step is to investigate the interaction of the pair fluxon-antifluxon. To fix the ideas let $k_1 > |k_2|$, $k_1 > 0$, $k_2 < 0 \Rightarrow k_1 \pm k_2 > 0$, $0 < c_1 < c_2 \Rightarrow e^{A_{12}} = \frac{(k_1-k_2)^2}{(k_1+k_2)^2} > 1$.

We repeat the previous considerations but in this situation $\frac{\partial \eta_1}{\partial x} > 0$, $\frac{\partial \eta_2}{\partial x} < 0$, i.e. η_1 is monotonically increasing in x and η_2 - decreasing in x.

$(\tilde{A})$ Fix $\eta_1 = const \Rightarrow x \sim c_1 t$ $\left(\begin{matrix} x \to +\infty \iff t \to \infty \\ x \to -\infty \iff t \to -\infty \end{matrix} \right)$ and $\eta_2 \sim k_2(c_1 - c_2)t$.

Therefore, $t \to -\infty \iff x \to -\infty \Rightarrow e^{\eta_2} \to 0 \Rightarrow u_2 \sim 4arctg(e^{\eta_1})$ (I). On the other hand, $t \to \infty \Rightarrow e^{\eta_2} \to \infty \Rightarrow u_2 \sim -2\pi + 4arctg(e^{A_{12}+\eta_1})$ (II).

$(\tilde{B})$ Fix now $\eta_2 = const \Rightarrow x \sim c_2 t$, $c_2 > 0$. Then $t \to -\infty \Rightarrow \eta_1 \sim k_1(c_2 - c_1)t$, $c_2 - c_1 > 0 \Rightarrow e^{\eta_1} \to 0 \Rightarrow u_2 \sim 4arctg(e^{\eta_2})$ (III). From $t \to \infty$ we have that $e^{\eta_1} \to \infty \Rightarrow u_2 \sim -2\pi + 4arctg(e^{A_{12}+\eta_2})$ (IV).

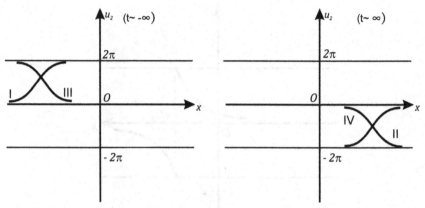

Fig. 6.7. I, II are fluxons (monotonically increasing), III-IV are antifluxons.

We shall study now the interaction fluxon-antifluxon for finite time t. Fix t. $\frac{\partial(\eta_1+\eta_2)}{\partial x} = k_1 + k_2 > 0$, $u_2 > 0$ for $x < x(t)$, $u_2 < 0$ for $x > x(t)$, $u_2(x) \to_{x\to-\infty} 2\pi$ as $e^{\eta_1} \to_{x\to-\infty} 0$, $e^{\eta_2} \to_{x\to-\infty} \infty$, $e^{\eta_1+\eta_2} \to_{x\to-\infty} 0$. Similarly, $u_2(x) \to_{x\to+\infty} -2\pi$. $u_2(x)$ is not monotonically increasing (see Fig. 6.8), u_2 has a jump at $x = x(t)$, $\lim_{x\to x(t)}\frac{\partial u_2}{\partial x} = 4(e^{\eta_1} + e^{\eta_2})^{-1}(k_1 + k_2)|_{x=x(t)} > 0$.

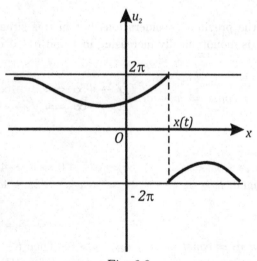

Fig. 6.8.

6.5.1 Rational solutions of some equations of mathematical physics

Below we propose several rational solutions of the Korteweg de Vries and Boussinesq equations [5]. They are deduced from the 1-soliton solutions of that equations:

$$u = 2(lnf_1)_{xx}, f_1 = 1 + e^\eta, \eta = Ax + Bt + \eta^{(0)}.$$

The constant term $\eta^{(0)}$ plays an important role in constructing of the rational solutions.

Consider the KdV equation. Its one soliton solution is given by the formula $u = \frac{k_1^2}{2} sech^2 \frac{k_1 x - k_1^3 t + \eta_1^{(0)}}{2}$. Let $e^{\eta_1^{(0)}} = -1$. Simple computations show that then $u = -\frac{k_1^2}{2} \frac{1}{sh^2 \frac{k_1 x - k_1^3 t}{2}} \to_{k_1 \to 0} -\frac{2}{x^2}$. u has a pole at $k_1 x - k_1^3 t = 0$ and $-\frac{2}{x^2}$ is a rational function.

The Boussinesq equation possesses the 1-soliton solution $u = 2(lnf_1)_{xx}$, $f_1 = 1 + e^{\eta_1}$, $\eta_1 = k_1(x + \sqrt{1 + k_1^2} t + \eta_1^{(0)})$. Taking again $e^{\eta_1^{(0)}} = -1$ we see that $f_1 = -k_1(x + \sqrt{1 + k_1^2} t) + O(k_1^2)$, $k_1 \to 0$, i.e. $f_1 \sim -k_1(x+t)$, $k_1 \to 0$. Consequently, $u \sim (\frac{2}{x+t+O(k_1)})_x \sim \frac{-2}{(x+t+O(k_1))^2}$ and $u \to_{k_1 \to 0} -\frac{2}{(x+t)^2}$; $-\frac{2}{(x+t)^2}$ is a rational function.

Consider the Eq. (6.1) - KdV. We constructed at the beginning of that chapter its two soliton solution $u = 2(logf_2)_{xx}$, where $f_2 = 1 + e^{\Theta_1} + e^{\Theta_2} + e^{\Theta_1 + \Theta_2 + b_{12}}$, $\Theta_i = k_i x - k_i^3 t + \delta_i$, $\delta_i = const$, $i = 1,2$, $e^{b_{12}} = (\frac{k_1 - k_2}{k_1 + k_2})^2$.

Exercise 3.

(a) Prove the formula

$$u = 2\frac{k_1^2 e^{\Theta_1} + k_2^2 e^{\Theta_2} + k_2^2 e^{2\Theta_1 + \Theta_2 + b_{12}} + k_1^2 e^{2\Theta_2 + \Theta_1 + b_{12}}}{(1 + e^{\Theta_1} + e^{\Theta_2} + e^{b_{12} + \Theta_1 + \Theta_2})^2} +$$
$$2\frac{e^{\Theta_1 + \Theta_2}(e^{b_{12}}(k_1 + k_2)^2 + (k_1 - k_2)^2)}{(1 + e^{\Theta_1} + e^{\Theta_2} + e^{b_{12} + \Theta_1 + \Theta_2})^2}.$$

(b) Prove that u can be written in the following trigonometric form:

$$u = \frac{pcosech^2\Theta_1 + qsech^2\Theta_2}{(rctgh\Theta_1 - stgh\Theta_2)^2},$$

where the parameters p, q, r, s describe the following line: $r = r$, $s = \frac{1}{\sqrt{C_1}} r$, $q = \frac{k_2^2}{2} \frac{C_1 - 1}{C_1} r^2$, $p = C_1 q = \frac{k_2^2}{2}(C_1 - 1)r^2$, $C_1 = \frac{k_1^2}{k_2^2} \neq 1$, $C_1 > 0$.

(c) Can you find other trigonometrical forms of u?

Answer: $sech^2\Theta_2 = \frac{1}{ch^2\Theta_2}$, $cosech^2\Theta_1 = \frac{1}{sh^2\Theta_1}$, $ch^2\Theta_2 = \frac{1 + ch2\Theta_2}{2}$, $sh^2\Theta_1 = \frac{ch2\Theta_2 - 1}{2}$, $ch\Theta_1 ch\Theta_2 = \frac{ch(\Theta_1 + \Theta_2) + ch(\Theta_1 - \Theta_2)}{2}$, $sh\Theta_1 sh\Theta_2 = \frac{ch(\Theta_1 + \Theta_2) - ch(\Theta_1 - \Theta_2)}{2}$.

Thus, $u = 2\dfrac{p+q+pch2\Theta_1+qch2\Theta_2}{[(r-s)ch(\Theta_1+\Theta_2)+(r+s)ch(\Theta_1-\Theta_2)]^2}$.

We proposed these three formulae as in many books on the subject the soliton solutions are written in the seemingly different forms (a), (b), (c) which are equivalent of course.

Chapter 7

Special type solutions of several evolution PDEs

7.1 Introduction

This chapter deals with several methods (elementary and not complicated) for finding of explicit solutions of some equations of mathematical physics, appearing in fluid mechanics, plasma physics, hydrodynamics, etc. More precisely, we study the Boussinesq, the Korteweg de Vries, the Kuznetsov-Zakharov, the Liouville, the Manakov, the cubic nonlinear Schrödinger, the Swift-Hohenberg equations as well as the nonlocal harmonic mode locking laser model and discuss the rogue waves. In this Introduction we shall briefly discuss the famous Boussinesq equation and its generalizations only. Special comments for other PDEs under consideration as Swift-Hohenberg and Kuznetsov-Zakharov will be proposed during the process of their investigation. From the point of view of the methods used here we can say that the Chapter is supplied with many examples and exercises illustrating the different approaches for solvability of the above mentioned equations. These approaches can be included into the more general scheme known as Method of simplest equation and its modifications.

Concerning the dispersive multidimensional Boussinesq equation [26] and some of its generalizations (for example the $B(m, n)$ Boussinesq equation, the Paradigm Boussinesq equation), they describe the formation of patterns in liquid drops, the vibrations of a single one-dimensional lattice and other effects from fluid mechanics. Many historical notes and interesting facts on the subject are available in the monographs [21], [138], [141].

There is a classical approach for studying of conservative mechanical systems having a degree of freedom equal to 1 (see for example [13]). Applying it or its modifications we find traveling wave solutions of many equations of mathematical physics (see Chapter III of [110]). A detailed study of the generalized $(N+1)$ dimensional dispersive Boussinesq equation from the point of view of the traveling wave solutions can be found in Section 3 of that Chapter. The solutions in question are written into integral form and it turns out that they develop cusp type singularities. In some special cases they are expressed by the Jacobi elliptic functions. The Weierstrass $\wp$ function can appear too. The main problem here is to find an appropriate Ansatz for the solutions.

7.2 Traveling waves. Method of the auxiliary solution (of the simplest equation)

Many physical phenomenons are attended with the processes of localized pattern formation in some open systems. Those structures while different in their origin can be described in many cases by the same or similar mathematical models. An example of such a model was proposed by Swift and Hohenberg in 1977 in [126] in investigating the hydrodynamics fluctuations at the convective instability. This model is based on the following semilinear fourth order evolution equation:

$$u_t + 2u_{xx} + u_{xxxx} = \alpha u + \beta u^2 - \gamma u^3. \tag{7.1}$$

Equation (7.1) appears for description of the average dislocation density under the formation of shear microbands in nanocrystaline materials, in finding the amplitude of the optical electric field inside the cavity containing a nonlinear medium, and for describing the patterns in thin vibrated granular layers. Different generalizations of Eq. (7.1) were proposed in [82], [83]. They concern a generalized dispersive Swift-Hohenberg equation of the form:

$$u_t + 2u_{xx} - \sigma u_{xxx} + u_{xxxx} = \alpha u - \gamma u^{m+1}, \tag{7.2}$$

where σ, α, γ are parameters and the integer $m \geq 1$. Thus, a dispersive term is added in Eq. (7.2).

We propose below a simple mathematical approach for constructing of traveling wave solutions $u(x, y) = \varphi(\alpha x + \beta y)$, $\xi = \alpha x + \beta y$ of the following fourth order semilinear PDE

$$Lu = \sum_{1 \leq i+k \leq 4} a_{ik} \partial_x^i \partial_y^k u(x, y) = \sum_{j=0}^{5} A_j u^j = F(u) \tag{7.3}$$

a_{ik}, A_j being constants.

Thus, $\varphi(\xi)$ should satisfy the ODE

$$L(\partial_\xi)u = \sum_{1 \le i+k \le 4} a_{ik}\alpha^i\beta^k\varphi^{(k+i)}(\xi) = \sum_{j=0}^{5} A_j\varphi^j(\xi) = F(\varphi). \qquad (7.4)$$

Our aim is to find a kink type solutions of Eq. (7.4), i.e. we are looking for a monotonically increasing (decreasing) smooth function φ and such that there exist $\varphi(\pm\infty)$, $\varphi(+\infty) = a$, $\varphi(-\infty) = b$, $a \ne b$, a, b-constants. Evidently, Eq. (7.4) should satisfy two conditions namely

$$\sum_{1 \le i+k \le 4} a_{ik}\alpha^i\beta^k\varphi^{(k+i)}(\pm\infty) = \sum_{j=0}^{5} A_j\varphi^j(\pm\infty). \qquad (7.5)$$

Certainly, it is assumed that $\varphi^{(k+i)}(\pm\infty) = 0$ exist.

To simplify the notations we shall write Eq. (7.4) in the form

$$k_1\varphi' + k_2\varphi'' + k_3\varphi''' + k_4\varphi^{iv} = \sum_0^5 A_j\varphi^j = F(\varphi). \qquad (7.6)$$

Then Eq. (7.5) takes the form

$$\sum_0^5 A_j a^j = 0, \sum_0^5 A_j b^j = 0. \qquad (7.7)$$

Below two cases will be considered separately:

(i) $ab \ne 0$;

(ii) $ab = 0$, $a^2 + b^2 > 0$.

In the case (ii) $A_0 = 0$ and $\sum_{j=1}^5 b^{j-1}A_j = 0$ for $a = 0$, $b \ne 0$. Therefore, $A_2, \ldots, A_5$ are free parameters, while $A_1 = -\sum_{j=2}^5 A_j b^{j-1}$.

In the case (i) A_0, A_3, A_4, A_5 are free parameters, while A_1, A_2 are expressed linearly by them. To simplify the things we shall take $a = 1$, $b = 0$ for (ii) and $a = 1$, $b = -1$ for (i). Kink type solutions are given by the formulas: (ii) $y(\xi) = \frac{1}{1+e^{-\xi}}$ where $y' = y-y^2$, $y = \frac{1}{2}(1+th(\frac{\xi}{2}))$, respectively $y(\xi) = tgh\xi$, where (i) $y' = 1 - y^2$. Having in mind that in case (ii) $y'' = y - 3y^2 + 2y^3$, $y''' = y - 7y^2 + 12y^3 - 6y^4$, $y^{iv} = y - 15y^2 + 50y^3 - 60y^4 + 24y^5$ and equalizing the coefficients at different powers of $\varphi \equiv y$ in the left and right-hand sides of Eq. (7.6) we obtain a linear algebraic system with respect to k_i, $1 \le i \le 4$, namely

$$k_1 + k_2 + k_3 + k_4 = A_1 \qquad (7.8)$$

$$-k_1 - 3k_2 - 7k_3 - 15k_4 = A_2$$

$$2k_2 + 12k_3 + 50k_4 = A_3$$

$$-6k_3 - 60k_4 = A_4$$

$$24k_4 = A_5$$

The system is containing 5 linear equations for the unknown quantities $k_1, \ldots, k_4$ and therefore is overdetermined. The 4×4 subsystem of Eq. (7.8) consisting of the last four equations is uniquely solvable with respect to k_i. These are the corresponding solutions

$$k_1 = -A_2 - \frac{3}{2}A_3 - \frac{11}{6}A_4 - \frac{25}{12}A_5 \qquad (7.9)$$

$$k_2 = \frac{A_3}{2} + A_4 + \frac{35A_5}{24}$$

$$k_3 = -\frac{A_4}{6} - \frac{5A_5}{12}$$

$$k_4 = \frac{A_5}{24}.$$

The first equation of Eq. (7.8) is satisfied iff

$$A_1 = k_1 + k_2 + k_3 + k_4 = -A_2 - A_3 - A_4 - A_5. \qquad (7.10)$$

Therefore, $A_1 = -A_2 - A_3 - A_4 - A_5$ is a necessary and sufficient condition for the existence of kink type solution of Eq. (7.6) in the case (ii), $a = 1$, $b = 0$.

Consider now the case (i), $a = -b = 1$. Then $y' = 1 - y^2$, $y'' = 2y(y^2 - 1)$, $y''' = -2 + 8y^2 - 6y^4$, $y^{iv} = 8y(2 - 5y^2 + 3y^4)$. Proceeding as in the previous case we obtain

$$k_1 = -A_2 - \frac{4}{3}A_4 \qquad (7.11)$$

$$k_2 = \frac{A_3}{2} + \frac{5A_5}{6}$$

$$k_3 = -\frac{A_4}{6}$$

$$k_4 = \frac{A_5}{24}.$$

The system analogue to Eq. (7.8) contains 6 equations with right-hand sides $A_0, A_1, \ldots, A_5$ and the unknown quantities are k_1, k_2, k_3, k_4. We have found them in a unique way by using the last 4 equations. As it concerns the first two equations they are

$$k_1 - 2k_3 = A_0 \qquad (7.12)$$

$$-2k_2 + 16k_4 = A_1.$$

Putting the expressions for k_i from Eq. (7.11) into Eq. (7.12) we obtain that Eq. (7.12) is equivalent to

$$A_1 = -A_3 - A_5 \qquad (7.13)$$

$$A_0 = -A_2 - A_4.$$

Consequently, according to Eq. (7.7) we have that Eq. (7.13) is a necessary and sufficient condition for the existence of kink type solution of Eq. (7.6) in the case (i), $a = -b = 1$. The hyperbolic functions (i), (ii) are the auxiliary functions in the case of equation Eq. (7.3) (Eq. (7.1), Eq. (7.2)).

Slight generalizations can be given by using the kink equations $y' = \alpha y + \beta y^2$, $\alpha > 0 > \beta$, respectively $y' = \alpha - \beta y^2$, $\alpha > 0$, $\beta > 0$. Other possible generalization is for right-hand side of Eq. (7.2), Eq. (7.3), Eq. (7.4), Eq. (7.6) given by $F(u) = A_1 u + A_{m+1} u^{m+1}$, $m \geq 1$.

Then we are looking for a kink solution having the form $\varphi = y^\alpha$, $\alpha > 0$, $y = \frac{1}{1+e^{-\xi}}$. One can easily see that: $A_1 + A_{m+1} = 0$ if φ satisfies the equation

$$\varphi' = \alpha y^{\alpha-1}(y - y^2) = \alpha y^\alpha - \alpha y^{\alpha+1} = \alpha y^\alpha(1 - y).$$

Moreover,

$$\varphi'' = \alpha^2 y^\alpha - \alpha(2\alpha + 1)y^{1+\alpha} - \alpha(\alpha + 1)y^{\alpha+2}, \text{ etc.}$$

Putting the expressions for $\varphi', \dots, \varphi^{iv}$ in Eq. (7.6) and having in mind that $F(\varphi) = A_1\varphi - A_2\varphi^{m+1} = A_1 y^\alpha - A_{m+1} y^{\alpha(m+1)}$ we can divide both sides of the equation by $y^\alpha > 0$. This way we obtain in the left-hand side of Eq. (7.6) a fourth order polynomial with respect to y, while in the right-hand side stands the expression $A_1 + A_{m+1} y^{\alpha m}$. Taking $\alpha m = 4$ we again equalize the coefficients at different powers of y (5 coefficients for 4 unknown quantities k_1, k_2, k_3, k_4 and in the right-hand side of the corresponding linear algebraic system we have $(A_{1+m}, 0, 0, 0, A_1)$. We omit the standard technical computations which are rather tiresome.

Another possibility is to look for a kink solution $\varphi = c_1 + c_2 y^\alpha$, $c_1, c_2 = const \neq 0$ etc.

Exercise 1. Find a soliton solution of the equation $k_1 y_1' + \dots + k_4 y_1^{iv} = A_1 y_1 - A_5 y_1^5$.

Hint. Put $y_1 = y e^{a\xi}$, $y' = y - y^2$, $y = \frac{1}{1+e^{-\xi}} \Rightarrow e^{-\xi} = \frac{1-y}{y}$. Therefore, $L_1(\partial_\xi)y = P_5(y) = y(A_1 - A_5 y^4 e^{4a\xi})$, $P_5(y)$ being fifth order polynomial of y. Taking $4a = -1$ we get: $P_5(y) = yA_1 - A_5(1 - y)y^4$ etc.

Remark 7.1. The change $\xi = x - ct$ converts each nonlinear PDE to an ODE of the form:

$$Q(u, u', u'', u''', \dots) = 0. \tag{7.14}$$

The standard tgh method introduces a new independent variable $z = tgh\mu\xi$, respectively

$$u = \sum_{i=0}^{n} a_i z^i. \tag{7.15}$$

Evidently, $z' = \mu(1-z^2)$, $z'' = -2\mu^2 z + 2\mu^2 z^3$ etc. Imitating the procedure applied to the Swift-Hohenberg equation we can find out solutions of several ODE of the form Eq. (7.14) via explicit formulas proposed by Eq. (7.15). One of the simplest PDE having traveling wave solutions is the famous Boussinesq equation

$$u_{tt} - u_{xx} - 3(u^2)_{xx} - u_{xxxx} = 0. \tag{7.16}$$

It can be easily converted to

$$(c^2 - 1)u - 3u^2 - u'' = 0. \tag{7.17}$$

The tgh method (i.e. the auxiliary function being of the type Eq. (7.15), see [136], [137]) admits the use of the change

$$u = a_0 + a_1 z + a_2 z^2, \tag{7.18}$$

the coefficients a_0, a_1, a_2 being unknown.

Substituting Eq. (7.18) into Eq. (7.17), collecting the coefficients of each power z^i, $0 \le i \le 4$ we get the nonlinear algebraic system with respect to a_0, a_1, a_2, μ, $a_2\mu \ne 0$:

$$\left|\begin{array}{l} 2a_2\mu^2 + 3a_0^2 + (1 - c^2)a_0 = 0 \\ -2a_1\mu^2 + 6a_0 a_1 + (1 - c^2)a_1 = 0 \\ -8a_2\mu^2 + 3a_1^2 + 6a_0 a_2 + (1 - c^2)a_2 = 0 \\ 2a_1\mu^2 + 6a_1 a_2 = 0 \\ 6a_2\mu^2 + 3a_2^2 = 0. \end{array}\right. \tag{7.19}$$

Thus, $\mu^2 = -\frac{1}{2}a_2$, $a_1(\mu^2 + 3a_2) = 0 \Rightarrow a_1 = 0$.

Therefore,

$$\left|\begin{array}{l} -8\mu^2 + 6a_0 + (1 - c^2) = 0 \\ 2a_2\mu^2 + 3a_0^2 + (1 - c^2)a_0 = 0, \end{array}\right. \tag{7.20}$$

i.e. $\mu^2 = -\frac{a_2}{2} > 0$, $a_2 = -\frac{3}{2}a_0 + \frac{c^2-1}{4}$,

$$a_{01,2} = \frac{c^2 - 1 \pm 2|c^2 - 1|}{6} = \frac{|c^2 - 1|}{6}[sgn(c^2 - 1) \pm 2].$$

Consequently, $a_2 = \mp\frac{|c^2-1|}{2}$ and we must take the sign "-". Then

$$a_0 = \frac{|c^2 - 1|}{6}[sgn(c^2 - 1) + 2], a_1 = 0, \tag{7.21}$$

$a_2 = -\frac{|c^2-1|}{2}$, $\mu = \frac{1}{2}\sqrt{|c^2 - 1|}$ etc.

Instead of tgh we can work with $sech\xi$, as $tgh^2 z = 1 - sech^2 z$ and in our case $u = a_0 + a_2 tgh^2(\mu\xi) = a_0 + a_2 - a_2 sech^2(\mu\xi)$. The solutions obtained by Eq. (7.18), Eq. (7.21) are solitons, of course.

We shall say several words about Boussinesq Paradigm equation [80] and shall formulate the Cauchy problem for it. Thus,

$$\frac{\partial^2 u}{\partial t^2} = \Delta u + \beta_1 \Delta \frac{\partial^2 u}{\partial t^2} - \beta^2 \Delta^2 u + \alpha_0 \Delta f(u) = 0, x \in \mathbf{R}^n, 0 \le t \le T < \infty$$
(7.22)

$$u(x,0) = u_0(x), u_t(x,0) = u_1(x)$$
$$u(x,t) \to 0, \Delta u(x,t) \to 0 \text{ as } |x| \to \infty.$$

Here u is a surface elevation, $f(u) = u^p$, $p \in \mathbf{N}$, $p \ge 2$, β_1, $\beta_2 \ge 0$, $\beta_1 + \beta_2 \ne 0$ are two dispersive coefficients and α_0 is an amplitude parameter. Equation (7.22) appears in the modeling of surface waves in shallow waters, in the theory of acoustic waves, in ion-sound waves, plasma and nonlinear lattice waves etc. [80]. Put in equation Eq. (7.22) $n = 1$, $f(u) = u^{m+1}$ and look for traveling waves solutions: $u(x,t) = \varphi(x - ct)$, $\xi = x - ct$, $c \ne 1$.

Then Eq. (7.22) takes the form (after two integrations in ξ and taking zero the integration constants):

$$(c^2 - 1)\varphi = (\beta_1 c^2 - \beta_2)\varphi'' + \alpha_0 \varphi^{m+1} = 0, m \ge 1,$$
(7.23)

i.e.

$$\varphi'' = A_1 \varphi + A_2 \varphi^{m+1}, A_1 > 0, A_2 < 0$$
(7.24)

(this case will be studied only).

To illustrate the *tghx* method we put

$$\varphi = k_2 (sech^2(k_1 x))^\alpha$$
(7.25)

with k_1, k_2, $\alpha > 0$ being unknown parameters.

Substituting Eq. (7.25) in Eq. (7.24) and having in mind that the function $sech^2(k_1 x) = 1 - tgh^2(k_1 x)$ and $z = tgh(k_1 x)$ satisfies the ODE $z' = k_1(1 - z^2)$ we get for $\alpha m = 1 \iff \alpha = \frac{1}{m}$:

$$-2k_1^2 \alpha + 2k_1^2 \alpha(1 + 2\alpha)z^2 = A_1 + A_2 k_2^m - A_2 k_2^m z^2,$$
(7.26)

i.e.

$$\begin{vmatrix} -2k_1^2 \alpha = A_1 + A_2 k_2^m \\ 2k_1^2 \alpha(1 + 2\alpha) = -A_2 k_2^m. \end{vmatrix}$$
(7.27)

So

$$k_2 = \left(\frac{-A_1(m+2)}{2A_2} \right)^{\frac{1}{m}}, k_1 = \pm \frac{1}{2}\sqrt{A_1 m}.$$

This way we conclude that

$$\varphi = \left[\frac{-A_1}{A_2} \frac{m+2}{2} sech^2 \left(\frac{1}{2} \sqrt{A_1} m \xi \right) \right]^{\frac{1}{m}} \qquad (7.28)$$

generates a traveling wave solution $u(t,x) = \varphi(x - ct)$ of Eq. (7.23), $n = 1$.

Here $A_1^{-1} = \frac{\beta_1 c^2 - \beta_2}{c^2 - 1} > 0$, $A_2^{-1} = \frac{\beta_1 c^2 - \beta_2}{-\alpha_0} < 0$. Evidently, $\varphi > 0$, $\varphi(\pm\infty) = 0$,

$$u(t,x) = \left[\frac{c^2 - 1}{\alpha_0} \frac{m+2}{2} sech^2 \frac{1}{2} \sqrt{\frac{(c^2 - 1)}{(\beta_1 c^2 - \beta_2)}} m(x - ct) \right]^{\frac{1}{m}}, m \geq 1. \quad (7.29)$$

The same formula Eq. (7.28) can be obtained by using the fact that for
$\begin{vmatrix} 0 < \varphi < 1 \\ \alpha > 0 \end{vmatrix}$: $\frac{d}{d\varphi} arcsech\varphi^\alpha = \frac{-\alpha}{\varphi\sqrt{1-\varphi^{2\alpha}}}$, i.e. $x = \int \frac{d\varphi}{\varphi\sqrt{1-\varphi^{2\alpha}}} = \frac{-arcsech\varphi^\alpha}{\alpha}$

and therefore $\varphi^\alpha = sech\alpha x \iff \varphi = (sech\alpha x)^{\frac{1}{\alpha}}$, $\alpha = \frac{m}{2}$ etc.

Exercise 2. Find traveling wave solutions of the following Paradigm Boussinesq equation:

$$\beta_2 u_{tt} - u_{xx} - \beta_1 u_{ttxx} + u_{xxxx} = (B_0 u^{p+1} + C_0 u^{2p+1})_{xx},$$

where $p > 0$, $\beta_1 > 0$, $\beta_2 > 0$ are some constants.

Hint. After two integrations by parts the solution $u(x,t) = \varphi(x - ct)$ satisfies the ODE

$$\varphi'' = A\varphi + B\varphi^{p+1} + C\varphi^{2p+1},$$

A, B, C being some constants, i.e $(\varphi')^2 = A\varphi^2 + B_1\varphi^{p+2} + C_1\varphi^{2p+2}$, where $B_1 = \frac{2B}{p+2}$, $C_1 = \frac{C}{p+1} \Rightarrow \xi + D = \pm \int \frac{d\varphi}{\varphi\sqrt{A+B_1\varphi^p+C_1\varphi^{2p}}} = \pm I(\varphi)$.

The change $z = \varphi^{-p}$ implies that $I(\varphi) = I(z^{-\frac{1}{p}}) = -\frac{1}{p} \int \frac{dz}{\sqrt{Az^2+B_1z+C_1}}$.
Depending on the sign of the discriminant one should consider the cases $Az^2 + B_1 z + C_1 \geq 0$ everywhere, respectively in intervals of the type $[z_1, z_2]$; $(-\infty, z_1]$, $(z_2, +\infty)$ if $A > 0$).

Answer. Let $A > 0$, $4AC_1 - B_1^2 > 0$. Then

$$\varphi = \left[\frac{1}{2A} \left(-B_1 \pm \sqrt{4AC_1 - B_1^2} \, sh \left(p\sqrt{A}(\xi + D) \right) \right) \right]^{-\frac{1}{p}},$$

$D = const$. The calculations are formal.

Below we propose as a Proposition several formulas useful in the applications to different equations arising in physics:

Proposition 7.1.

1) Consider the ODE

$$y'' = Ay + By^3. \qquad (7.30)$$

Then:

(a) For $A < 0$ and $B > 0$ it has the solutions $y = \pm\sqrt{\frac{-2A}{B}}\sec(\sqrt{-A}x)$,
$\sec z = \frac{1}{\cos z}$, $y = \sqrt{\frac{-A}{B}}tgh(x\sqrt{-\frac{A}{2}})$;

(b) For $A > 0$, $B < 0$ it has the solution $y = \pm\sqrt{\frac{-2A}{B}}sech(\sqrt{A}x)$;

(c) $y = \pm\sqrt{\frac{AC_3}{C_2 B}}sn(\sqrt{\frac{A}{C_2}}x, k)$ for $C_2 = -(1 + k^2)$, $C_3 = 2k^2$, $y = \pm\sqrt{\frac{AC_3}{C_2 B}}cn(\sqrt{\frac{A}{C_2}}x, k)$ for $C_2 = 2k^2 - 1$, $C_3 = -2k^2$, $0 < k^2 < 1$, $\frac{A}{C_2} > 0$, $\frac{C_3}{B} > 0$ are solutions of Eq. (7.30);

(d) For $A > 0$, $B > 0$ it has the solution $y = \pm\sqrt{\frac{2A}{B}}cosech\sqrt{A}x$,
$cosech x = \frac{2}{e^x - e^{-x}} = \frac{1}{shx}$.

2). Consider the Bernoulli ODE

$$y' = Ay + By^k, k \geq 2. \tag{7.31}$$

Then

(e) $y(x) = \left(\frac{Ae^{A(k-1)x}}{1 - Be^{A(k-1)x}}\right)^{\frac{1}{k-1}}$ satisfies the equation Eq. (7.31) for $A > 0$, $B < 0$;

(f) $y(x) = \left(-\frac{Ae^{A(k-1)x}}{1 + Be^{A(k-1)x}}\right)^{\frac{1}{k-1}}$ is a solution of that equation for $A < 0$, $B > 0$.

Our next step is an initial investigation of the Manakov system [95]:

$$-iA_0\frac{\partial u_1}{\partial t} + A_1\frac{\partial^2 u_1}{\partial x^2} + A_2(|u_1|^2 + |u_2|^2)u_1 + \lambda_1 u_1 = 0 \tag{7.32}$$

$$-iB_0\frac{\partial u_2}{\partial t} + B_1\frac{\partial^2 u_2}{\partial x^2} + B_2(|u_1|^2 + |u_2|^2)u_2 + \lambda_2 u_2 = 0,$$

where $u_1 = u_1(t, x)$, $u_2 = u_2(t, x)$, the coefficients being real-valued. It appears in the theory of 2D stationary self-focusing of electromagnetic waves, in the theory of elementary particles, etc. We are looking for solutions of Eq. (7.32) having the form

$$u_1 = k_1(x)e^{i(\frac{\varepsilon_1}{A_0}t + Q_1(x))}, \varepsilon_1 \in \mathbf{R}^1 \setminus 0 \tag{7.33}$$

$$u_2 = k_2(x)e^{i(\frac{\varepsilon_2}{B_0}t + Q_2(x))}, \varepsilon_2 \in \mathbf{R}^1 \setminus 0$$

and we suppose that k_1, k_2, Q_1, Q_2 are real-valued smooth functions. Putting Eq. (7.33) into Eq. (7.32) we get:

$$A_1(k_1'' - k_1(Q_1')^2) + A_2(k_1^2 + k_2^2)k_1 + (\lambda_1 + \varepsilon_1)k_1 + \tag{7.34}$$
$$A_1 i(2k_1'Q_1' + k_1Q_1'') = 0$$
$$B_1(k_2'' - k_2(Q_2')^2) + B_2(k_1^2 + k_2^2)k_2 + (\lambda_2 + \varepsilon_2)k_2 +$$
$$B_1 i(2k_2'Q_2' + k_2Q_2'') = 0.$$

It is reasonable to find out Q_1, Q_2 in such way that $2k_1'Q_1' = -k_1Q_1''$, $2k_2'Q_2' = -k_2Q_2''$.

Thus, $Q_1 = C_1 \int \frac{dx}{k_1^2}$, $Q_2 = C_2 \int \frac{dx}{k_2^2}$. So,

$$A_1(k_1^3 k_1'' - C_1^2) + A_2(k_1^2 + k_2^2)k_1^4 + (\lambda_1 + \varepsilon_1)k_1^4 = 0 \qquad (7.35)$$
$$B_1(k_2^3 k_2'' - C_2^2) + B_2(k_1^2 + k_2^2)k_2^4 + (\lambda_2 + \varepsilon_2)k_2^4 = 0$$

Assuming that $k_1(x_0) \neq 0$, $k_2(x_0) \neq 0$ we can find a local solution of Eq. (7.35) near x_0. Assume now that $C_1 = C_2 = 0$ and $k_2 = \mu k_1$ for some $\mu \neq 0$. Therefore, Eq. (7.35) can be written as

$$A_1 k_1'' + A_2(k_1^2 + k_2^2)k_1 + (\lambda_1 + \varepsilon_1)k_1 = 0 \qquad (7.36)$$
$$B_1 k_1'' + B_2(1 + \mu^2)k_1^3 + (\lambda_2 + \varepsilon_2)k_1 = 0.$$

The equations participating in Eq. (7.36) are equivalent if $\frac{A_1}{B_1} = \frac{A_2}{B_2} = \frac{\lambda_1 + \varepsilon_1}{\lambda_2 + \varepsilon_2} \neq 0$, as $(k_1^2 + k_2^2)k_1 = (1 + \mu^2)k_1^3$.

This way we come to the following

Proposition 7.2.

Consider the Manakov system Eq. (7.32) with nonzero real coefficients and suppose that for some $\varepsilon_1 \neq 0$, $\varepsilon_2 \neq 0$; $\mu = \frac{A_1}{B_1} = \frac{A_2}{B_2} = \frac{\lambda_1 + \varepsilon_1}{\lambda_2 + \varepsilon_2} \neq 0$.

Then for that $\mu \in \mathbf{R}^1 \setminus 0$ the system Eq. (7.32) possesses the solution $u_{1\mu}$, $u_{2\mu}$, where $u_{1\mu} = k_{1\mu}(x)e^{i\frac{\varepsilon_1}{A_0}t}$, $u_{2\mu} = \mu k_{1\mu}(x)e^{i\frac{\varepsilon_2}{B_0}t}$ and $k_{1\mu}$ satisfies the equation $A_1 k_1'' + A_2(1 + \mu^2)k_1^3 + (\lambda_1 + \varepsilon_1)k_1 = 0$.

Remark 7.2. Taking $B = -\frac{A_2(1+\mu^2)}{A_1}$, $A = -\frac{\lambda_1 + \varepsilon_1}{A_1}$ in the previous ODE and applying Proposition 7.1 (a), (b), (c), (d) we obtain into explicit form kink type solution (in the physical terminology it is called dark soliton solution), soliton type solution (called bright soliton solution), periodic solution and exploding for finite x_0 solution. Generalizations of Eq. (7.25) are proposed in [81].

Consider now the cubic Schrödinger operator

$$iu_t + u_{xx} + A|u|^2 u = 0, A \in \mathbf{R}^1 \setminus 0. \qquad (7.37)$$

Again we look for

$$u(t,x) = k(x)e^{-i\varepsilon t + iQ}, \varepsilon \in \mathbf{R}^1 \setminus 0,$$

Q, k — being smooth real valued functions. As in the case of Eq. (7.32) we take $Q = C \int \frac{dx}{k^2(x)}$ and therefore

$$\varepsilon k^4 + (k^3 k'' - C^2) + Ak^6 = 0, \qquad (7.38)$$

i.e.

$$x = \pm \int \frac{kdk}{\sqrt{-(C^2 + \frac{A}{2}k^6 + \varepsilon k^4)}} = \pm \frac{1}{2} \int \frac{d(k^2)}{\sqrt{-(C^2 + \frac{A}{2}(k^2)^3 + \varepsilon(k^2)^2)}}$$

etc.

For the sake of simplicity put $C = 0 (\Rightarrow Q = 0)$. Then Eq. (7.38) takes the form

$$k'' = -Ak^3 - \varepsilon k. \tag{7.39}$$

According to Proposition 7.1 for $A < 0$ and $\varepsilon > 0$: $k_1(x) = \pm\sqrt{\frac{2\varepsilon}{-A}} sec(\sqrt{\varepsilon}x)$ or $k_2 = \sqrt{\frac{-\varepsilon}{A}} tgh(x\sqrt{\frac{\varepsilon}{2}})$, while for $\varepsilon < 0$, $A > 0$ $k(x) = \pm\sqrt{\frac{2\varepsilon}{-A}} sech(\sqrt{-\varepsilon}x)$. It is interesting to point out that for $A < 0$, $\varepsilon < 0 \Rightarrow k(x) = \sqrt{\frac{2\varepsilon}{A}} cosech(\sqrt{-\varepsilon}x)$, i.e. the solution blows up for $x = 0$, existing as a smooth function for $x < 0$ and for $x > 0$. In other words, dark solitons exist for $A < 0$ and bright exist for $A > 0$. Certainly, $u = e^{i(-\varepsilon t + vx)}$ is a traveling wave solution of Eq. (7.37) for $A + \varepsilon \geq 0$, $v^2 = A + \varepsilon$.

Having constructed the above mentioned solutions of Eq. (7.37) we can try to enlarge the class of special solutions of the nonlinear cubic Schrödinger operators. To do this we shall look for a solution of Eq. (7.37) having the form

$$u(t, x) = k(a_1 t + b_1 x)e^{i(a_2 t + b_2 x)}, a_1, \ldots, b_2 \in \mathbf{R}^1,$$

$a_1^2 + b_1^2 > 0$, $a_2^2 + b_2^2 > 0$ and $\xi = a_1 t + b_1 x$; $k(\xi)$ is real-valued function.

It is easy to see that then

$$k'(a_1 + 2b_1 b_2) = 0 \tag{7.40}$$

$$b_1^2 k'' - (a_2 + b_2^2)k + Ak^3 = 0.$$

We suppose that

$$a_1 + 2b_1 b_2 = 0 (\Rightarrow b_1 \neq 0). \tag{7.41}$$

Then k satisfies the equation

$$k'' = \frac{a_2 + b_2^2}{b_1^2} k - \frac{A}{b_1^2} k^3. \tag{7.42}$$

As we studied the case $b_2 = 0 = a_1$ we shall assume that

$$b_1 b_2 \neq 0, a_1 = -2b_1 b_2 \neq 0. \tag{7.43}$$

Proposition 7.3.

The equation Eq. (7.37) possesses a solution of the form $u = k(a_1 t + b_1 x)e^{i(a_2 t + b_2 x)}$ under the condition Eq. (7.43). It is a kink for $a_2 + b_2^2 < 0$ (i.e. $a_2 < 0$) and $A < 0$, soliton for $a_2 + b_2^2 > 0$, $A > 0$ and explodes for $a_2 + b_2^2 > 0$, $A < 0$. Periodic solutions can exist too for a special choice of the parameters A, a_1, a_2, b_1, b_2.

We propose now several exercises. This way the reader will get the ability to solve on his own different equations of mathematical physics.

The so-called Defocusing nonlinear Schrödinger operator is of interest to physics as then catastrophic waves can appear (see Section 4). Because of this reason we propose the following exercise.

Exercise 3. Consider the Defocusing nonlinear Schrödinger operator

$$iu_t + u_{xx} - |u|^2 u = 0,$$

i.e. after the change $t \to -t$ it becomes

$$iu_t = u_{xx} - |u|^2 u.$$

Prove that it possesses singular solutions with poles at some curves. When the solutions are not traveling waves?

Hint. Look for a solution of the type

$$u = e^{i(Ax+Bt)}\varphi(x - ct),$$

where $AB \neq 0$ are real constants, $c \in \mathbf{R}^1 \setminus 0$ and φ is a real valued function. The exercise is reduced to $2A = -c$ and to the solvability of the following ODE:

$$\varphi'' = (A^2 - B)\varphi + \varphi^3, \varphi = \varphi(\xi), \xi = x - ct.$$

Answer. If $B = A^2$ then $u = e^{iA(x+At)} \frac{\sqrt{2}}{x+2At}$; u is not a traveling wave.

If $A^2 > B$ then $u = e^{i(Ax+Bt)}\sqrt{2(A^2 - B)}cosech(\sqrt{A^2 - B}(x + 2At))$, u is not a traveling wave.

If $0 < A^2 < B$ then $u = e^{i(Ax+Bt)}\sqrt{2(B - A^2)}sec(\sqrt{B - A^2}(x + 2At))$, u is a traveling wave if and only if $B = 2A^2$. The phase functions are smooth, while the amplitudes vanish for $x + 2At = 0$, respectively for $x + 2At = \frac{\pi}{2\sqrt{B-A^2}}$.

Exercise 4. Consider the Derivative nonlinear Schrödinger "-" equation

$$iu_t + u_{xx} - i(|u|^2 u)_x = 0.$$

Prove that it possesses a smooth solution which is not a traveling wave.

Hint. Use the Ansatz $u = k\frac{(1+i\xi)}{(1-i\xi)^2}e^{i\eta}$, where $\xi = Ax+Bt$, $\eta = Mx+Nt$ and A, B, M, N, k are some real constants, different from 0.

Answer. $A = \frac{k^2}{4}$, $B = \frac{k^4}{16}$, $M = -\frac{k^2}{8}$, $N = -\frac{k^4}{64}$, $k \neq 0$. The solution is $u = k\frac{(1+i\frac{k^2}{16}(4x+k^2t))}{(1-i\frac{k^2}{16}(4x+k^2t))^2}e^{-i\frac{k^2}{64}(8x+k^2t)}$, $M > 0$.

Exercise 5. Consider the Derivative nonlinear Schrödinger "+" equation

$$iu_t + u_{xx} + i(|u|^2 u)_x = 0$$

and find a smooth solution which is not a traveling wave.

Exercise 6. Find traveling wave solutions of the derivative nonlinear Schrödinger " $\pm$ " equations.

Answer. $u = \varphi(x + ct)$, where $\varphi = Re^{ik\xi}$, $\xi = x + ct$, $c + k \pm R^2 = 0$; $c, k, R \in \mathbf{R}^1$.

In an optical fiber laser cavity the underlying wave behavior is governed by the nonlinear Schrödinger equation [125]. The latter is derived from a high-frequency asymptotic expansion of Maxwell's equation by using the paraxial approximation, rotating phase approximation, and quasi-monochromatic wave assumption in succession. The dominant physical effects are then the chromatic dispersion and a weak Kerr nonlinearity given by

$$i\frac{\partial Q}{\partial z} + \frac{1}{2}\frac{\partial^2 Q}{\partial t^2} + |Q|^2 Q = 0. \qquad (7.44)$$

Here Q is the electric field envelope, z represents the propagation distance in the fiber, and t is the time in the rest frame of a moving pulse. The function $Q(z, t) = secht e^{i\frac{z}{2}}$ satisfies Eq. (7.44).

Another mode-locked laser model is the harmonic mode-locking one. It can be described by the following nonlocal PDE (see [23]):

$$i\frac{\partial Q}{\partial z} + \frac{1}{2}\frac{\partial^2 Q}{\partial t^2} + |Q|^2 Q + iG\left(1 - 2\frac{\int_{-\infty}^t |Q|^2 d\tau}{\int_{-\infty}^\infty |Q|^2 d\tau}\right)Q = 0, \qquad (7.45)$$

where G measures the strength of the gain saturation across the pulse profile.

More details on the subject can be found in the paper [86].

Exercise 7. Find a solution of Eq. (7.45) having the form $Q = \varphi(t - Gz)e^{i\frac{z}{2}}$, $G = const \neq 0$, G-real-valued and $\varphi \in C^2(\mathbf{R}_\xi^1)$ is a real-valued function to be determined.

Hint. Substitute Q in Eq. (7.45) and put $\xi = t - Gz \in \mathbf{R}^1$.

This way one sees that φ satisfies the over determined system

$$\left|\begin{array}{l} \frac{\varphi''}{2} - \frac{1}{2}\varphi + \varphi^3 = 0 \\[2mm] \varphi' - \left(1 - 2\frac{\int_{-\infty}^\xi \varphi^2(\lambda)d\lambda}{\int_{-\infty}^\infty \varphi^2(\lambda)d\lambda}\right)\varphi = 0. \end{array}\right. \qquad (7.46)$$

From the second equation of Eq. (7.46) it follows that $\varphi''\varphi - (\varphi')^2 + \frac{2}{A}\varphi^4 = 0$, where $A = \int_{-\infty}^\infty \varphi^2(\lambda)d\lambda > 0$.

By the standard change $\varphi' = p(\varphi)$ one obtains that $p^2 = q$ is a solution of the linear ODE

$$\frac{dq}{d\varphi} - \frac{2}{\varphi}Aq = -\frac{4}{A}\varphi^3,$$

i.e.

$$q(\varphi) = \varphi^2 \left(C - \frac{2}{A}\varphi^2 \right), C > 0.$$

Therefore, $\varphi'(\xi) = \pm(C - \frac{2}{A}\varphi^2)^{1/2} \Rightarrow \xi = \pm\frac{1}{\sqrt{C}}arcsech\frac{\varphi}{\sqrt{\frac{AC}{2}}} \Rightarrow \varphi = \sqrt{\frac{AC}{2}}sech(\sqrt{C}\xi)$.

As $A = \int_{-\infty}^{\infty} \varphi^2(\xi)d\xi$ we get that $A = A\sqrt{C} \Rightarrow C = 1 \Rightarrow \varphi = \sqrt{\frac{A}{2}}sech\xi$. Taking $A = 2$ we see that $\varphi = sech\xi$ is a solution of the first equation of Eq. (7.46).

Answer. $u(z,t) = sech(t - Gz)e^{i\frac{z}{2}}$ satisfies the nonlocal integrodifferential equation Eq. (7.45).

As it is known from [76], [104], the pulse evolution in the actively mode-locked, optical fiber laser system is given by the equation

$$i\frac{\partial Q}{\partial z} + \frac{1}{2}\frac{\partial^2 Q}{\partial t^2} + Q|Q|^2 - ig\left(1 + \tau\frac{\partial^2}{\partial t^2}\right)Q + iM(\Gamma - cn^2(t,k))Q = 0, \quad (7.47)$$

where the constant τ controls the gain bandwidth, the parameters M, Γ measure the normalized strength of the active modulation element, which is responsible for generating the mode-locked pulse stream. The gain in the fiber is incorporated through the dimensionless parameter $g = const$ and $cn(t,k)$, $0 < k \leq 1$ is the Jacobi elliptic cosine function.

Exercise 8. Show that Eq. (7.47) possesses two families of solutions having the following explicit form:

a). $Q(z,t) = kcn(t,k)e^{-i(\frac{1}{2}-k^2)z}$, where $M = -2k^2g\tau$, $\Gamma = \frac{(2k^2-\frac{1}{\tau}-1)}{2k^2}$.

This solution branch represents a periodic train of pulses where adjacent are separated by a node. This nodal separation plays crucial role in the stability of the pulse trains.

b). $Q(z,t) = dn(t,k)e^{-i(\frac{k^2}{2}-1)z}$, where $M = -2k^2g\tau$, $\Gamma = \frac{k^2}{2}(2\tau(1 - \frac{1}{k^2}) + \frac{1}{k^2} + \frac{\tau(2-k^2)}{k^2})$ and $dn(t,k)$ is the Jacobi elliptic delta function.

This solution branch represents a periodic train of pulses where adjacent pulses are not separated by a node. As it is shown in [104] such solutions are unstable. The reduced separation of the neighbouring pulses causes the pulses to destabilize and evolve to a chaotic pulse train.

The following ansatz to the approximation solution of Eq. (7.47) was proposed in [27], [86], [113]:

$$Q(z,t) = A(z)cn(t,k) + B(z)dn(t,k).$$

In [86] and for $\tau \ll 1$ it is shown that the amplitudes $A(z)$, $B(z)$ depending on the propagation distance only should satisfy approximately (we do not enter into details here because of the lack of space) the ODE system:

$$i\frac{dA}{dz} + \frac{2k^2-1}{2}A + 2(1-k^2)|B|^2A = 0$$

$$i\frac{dB}{dz} + \frac{2-k^2}{2}B + 2\left(1-\frac{1}{k^2}\right)|A|^2B, 0 < k < 1.$$

Exercise 9. Solve the above written complex valued system of ODE. Certainly, only some of the infinitely many solutions of the system can have physical sense. Their special properties are beyond the scope of our book.

After so many exercises we shall formulate now the method of the simplest equation for obtaining of exact traveling wave solutions of some nonlinear PDE [134]. So we consider the ODE Eq. (7.14) and look for a solution having the form

$$u = \sum_{\mu=0}^{\nu} \Theta_\mu[y(\xi)]^\mu, \tag{7.48}$$

where $\nu > 0$ and Θ_μ are unknown quantities (constants).

The simplest equation is some ordinary differential equation of order less than the order of Eq. (7.14) and we know at least one exact special solution $y(\xi)$ of it. In many cases it is convenient to use as simplest equations Bernoulli Eq. (7.31) or Ricatti type ODE.

Remark 7.3. Let

$$\frac{dy}{d\xi} = \sum_{\mu=0}^{\beta} \gamma_\mu y^\mu, \tag{7.49}$$

where γ_μ are real parameters.

Certainly, $\xi + c = \int \frac{dy}{\sum_0^\beta \gamma_\mu y^\mu}$.

We search for a solution of Eq. (7.14) of the kind Eq. (7.48), Eq. (7.49). The substitution of Eq. (7.48) in Eq. (7.14) and Eq. (7.49) lead to the following polynomial equation:

$$P(y) = k_0 y^r + k_1 y^{r-1} + \ldots + k_r \equiv 0. \tag{7.50}$$

Here r is some integer and the coefficients k_i depend on nonlinear way on the parameters Θ_μ and γ_ν. Equation (7.50) is equivalent to the fulfilment of a system on nonlinear algebraic equations:

$$k_i = 0, i = 0, \ldots, r. \tag{7.51}$$

Finding (if possible) the parameters we are able to construct u.

We shall illustrate the above proposed method by a simple example-the classic Korteweg-de Vries (KdV) equation:

$$p_1 \frac{\partial u}{\partial t} + p_2 u \frac{\partial u}{\partial x} + p_3 \frac{\partial^3 u}{\partial x^3} = 0, p_1 p_2 p_3 \neq 0. \tag{7.52}$$

Put $u = u(x - vt)$, $\xi = x - vt$, i.e. $u = u(\xi)$ and search for

$$u = \Theta_0 + \Theta_1 y + \Theta_2 y^2, \tag{7.53}$$

where the simplest equation is

$$y' = Ay + By^2, AB \neq 0. \tag{7.54}$$

There are no difficulties to verify that

$$u^{(l)} = y P_{l+1}(y), l \in \mathbf{N}$$

and P_{l+1} is polynomial of order $l + 1$. Therefore, $u(\xi)$ satisfies

$$-vp_1 u' + up_2 u' + p_3 u''' = 0 \tag{7.55}$$

and after the substitution of Eq. (7.53) in Eq. (7.55) we get the following algebraic equation: $y(-vp_1 P_2(y) + p_2 P_2(y)(\Theta_0 + \Theta_1 y + \Theta_2 y^2) + p_3 P_4(y)) \equiv 0$, i.e.

$$P_2(-vp_1 + p_2(\Theta_0 + \Theta_1 y + \Theta_2 y^2)) + p_3 P_4(y) \equiv 0. \tag{7.56}$$

Consequently, the coefficients of the polynomial Eq. (7.56) must vanish and we obtain a nonlinear algebraic system with respect to Θ_0, Θ_1, Θ_2, A, B consisting of 5 equations. One can check that

$$\Theta_0 = \frac{p_1 v - p_3 A^2}{p_2}, \Theta_1 = \frac{-12 p_3 AB}{p_2}, \Theta_2 = -12 \frac{p_3 B^2}{p_2}$$

satisfy that nonlinear algebraic system.

The corresponding solution of Eq. (7.55) for $B < 0$, $A > 0$ is given according to Propositions 7.1 case 2) by the formula

$$u(\xi) = \frac{p_1 v - p_3 A^2}{p_2} - 12 \frac{p_3 A^2 B}{p_2} \left(\frac{e^{A\xi}}{1 - Be^{A\xi}} \right) \left(1 + B \frac{e^{A\xi}}{1 - Be^{A\xi}} \right). \tag{7.57}$$

Similar formula holds for $B > 0$, $A < 0$ [134].

Remark 7.4. Consider the equation $(y')^2 = a_0 + a_1 y^2 + a_2 y^4$, $a_0 a_1 a_2 \neq 0$. In [55] and [57] tables containing the expressions of the solutions of that ODE via the Jacobi elliptic functions are given. Put $z = \sqrt{A + By^2}$, where either $A > 0$, $B > 0$ or $A > 0$, $B < 0$ and $A > |B|$. Then

$$(z')^2 = \frac{1}{Bz^2}(B_0 + B_4 z^4 + B_6 z^6),$$

where $\lambda = \frac{B}{A}$ satisfies the equation

$$(*) \qquad a_0\lambda^2 - 2a_1\lambda + 3a_2 = 0$$

and $B_0 = A^3(-a_0\lambda^2 + a_1\lambda - a_2)$, $B_4 = A(a_1\lambda - 3a_2)$, $B_6 = a_2$. We assume that $a_1^2 - 3a_2a_0 > 0$ and for the sake of simplicity $a_0 > 0$. One can easily see that $(*)$ possesses at least one positive root λ iff the condition $a_1 < 0$, $a_2 > 0$ does not hold. On the other hand, $(*)$ possesses at least one root in the interval $(-1, 0)$ iff either

(i) $a_1 < 0$, $a_2 > 0$ and if $a_0 < -a_1$ then $a_0 + 3a_2 + 2a_1 < 0$

or

(ii) $a_2 < 0$, $a_0 > -a_1$, $a_0 + 3a_2 + 2a_1 > 0$.

This way we can solve explicitly the equation Eq. (7.38) in the case $C \neq 0$.

We shall say several words about the ODE

$$(w')^2 = Q_3(w), w = w(x), Q = a_0w^3 + a_1w^2 + a_2w + a_3, a_0 > 0,$$

where a_i are real constants. Then its solutions can be written in the form $w(x) = \wp(\frac{\sqrt{a_0}}{2}x, g_1, g_2) - \frac{a_1}{3a_0}$, the invariants g_1, g_2 being expressed by a_0, a_1, a_2, a_3. $\wp$ is the famous Weierstrass function (see for example [139]).

Consider the cubic polynomial $P_3(y) = 4y^3 - g_2z - g_3$ with real-valued coefficients g_2, g_3 and such that $\Delta = g_2^3 - 27g_3^2 \neq 0$. Moreover, $\Delta > 0$ implies that its roots $e_1 > e_2 > e_3$ are real and distinct: $\Delta = 16(e_1 - e_2)^2(e_1 - e_3)^2(e_2 - e_3)^2$, while $\Delta < 0$ implies that $P_3(y)$ possesses only one real root. Then the even Weierstrass function $\wp(x, g_2, g_3)$ satisfies the ODE $(\wp')^2 = 4\wp^3 - g_2\wp - g_2 = P_3(\wp)$ in the domains of x, where $P_3(\wp) \geq 0$, $\int_{\wp(x)}^{\infty} \frac{d\lambda}{\sqrt{P_3(\lambda)}} = x$. $\wp$ can be meromorphically and periodically prolonged in $\mathbf{C}^1 \setminus 0$ with poles $0, 2m\omega + 2n\omega'$, $m, n \in \mathbf{Z}$ and periods 2ω, $2\omega'$, $Im\frac{\omega'}{\omega} > 0$. For $x \in \mathbf{R}^1$ the coefficients of Taylor expansion of $\wp(x)$ are real:

$$\wp(x) = \frac{1}{x^2} + \frac{g_2}{20}x^2 + \frac{g_3}{28}x^4 + \dots.$$

Moreover, the period $\omega = \int_{e_1}^{\infty} \frac{d\lambda}{\sqrt{4\lambda^3 - g_2\lambda - g_3}}$, $\omega' = i\int_{-e_3}^{\infty} \frac{d\lambda}{\sqrt{4\lambda^3 - g_2\lambda - g_3}}$ for $\Delta > 0$.

If $\Delta > 0$ then $\wp(x, g_2, g_3) = e_3 + \frac{e_1 - e_3}{sn^2(\sqrt{e_1 - e_3}x, r)}$, $r = \frac{e_2 - e_3}{e_1 - e_3}$ (see [139]).

Exercise 10. Find out conditions on the coefficients A_1, A_2, λ_1, $B_1 \neq 0$, $B_2 \neq 0$, λ_2 of the system Eq. (7.35) under which it possesses solutions of the form $k_1^2 = C_1\wp^2 + C_2$, $C_1 \neq 0$ and $k_2^2 = C_3\wp^2 + C_4$, $C_3 \neq 0$.

Hint. k_1^2, k_2^2 are polynomials of $\wp$ of second degree. So the maximal power of $\wp$ in Eq. (7.35) is 6.

As it concerns $k_1^3 k_1^{''}$ it gives raise to a polynomial of fifth degree in $\wp$. As the coefficients of that polynomial of $\wp$, depending on A_1, A_2, λ_1, B_1, B_2, λ_2, $C_1 - C_4$, should vanish we assume at first that $C_1 + C_3 = 0$.

This way for the parameters one gets 2.6 algebraic nonlinear equations.

Consider also Eq. (7.32) with special potential terms, i.e. instead of λ_1 put $\lambda_1 \wp^2$ and instead of $\lambda_2 \to \lambda_2 \wp^2$, $\lambda_1 \lambda_2 \neq 0$. Then $A_2(C_1 + C_3) + \lambda_1 = 0$, $B_2(C_1 + C_3) + \lambda_2 = 0 \iff C_1 + C_3 = -\frac{\lambda_1}{A_2} = -\frac{\lambda_2}{B_2}$ etc.

In [6] the following formula is given

$$\int \frac{d\lambda}{\wp(\lambda) - \wp(\nu)} = \frac{1}{\wp'(\nu)} \left[2\lambda\zeta(\nu) + log\frac{\sigma(\lambda - \nu)}{\sigma(\lambda + \nu)} \right], \qquad (7.58)$$

where $\wp$, ζ, σ are standard Weierstrass functions. An introduction to their theory can be found in [6], [8], [33], [139]. Working absolutely formally in $\mathbf{C}^1$ we assume that $\frac{C_1}{C_2} < 0$ and one can find $\nu_1 \in \mathbf{C}^1$ and such that $\wp(\nu_1) = \sqrt{-C_2/C_1}$. Then

$$\int \frac{d\lambda}{k_1^2} = \frac{1}{2C_1\wp(\nu_1)} \int \left[\frac{1}{\wp(\lambda) - \wp(\nu_1)} - \frac{1}{\wp(\lambda) + \wp(\nu_1)} \right] d\lambda.$$

Under the additional assumption that there exists $\nu_2 \in \mathbf{C}^1$ and such that $\wp(\nu_2) = -\wp(\nu_1)$ we conclude that we can calculate explicitly the function Q_1 from Eq. (7.33)–Eq. (7.34).

Zakharov and Kuznetsov proposed in 1974 in [142] a nonlinear PDE in studying of the ion-acoustic waves in magnetized plasmas. In 2007 Moslem et al. [99] (see also [20]) derived the quantum Zakharov-Kuznetsov equation for electron-ion quantum plasmas. The corresponding equation can be written as

$$\frac{\partial u}{\partial t} + Au\frac{\partial u}{\partial z} + B\frac{\partial^3 u}{\partial z^3} + C\frac{\partial}{\partial z}\left(\frac{\partial^2}{\partial x^2} + \frac{\partial^2}{\partial y^2} \right) u = 0. \qquad (7.59)$$

Looking for traveling wave solution $u = u(\alpha x + \beta y + \gamma z - \nu t)$, $\xi = \alpha x + \beta y + \gamma z - \nu t$, we come to the following ODE:

$$Vu - A\frac{u^2}{2} - Eu^{''} = 0, E = const, V = const. \qquad (7.60)$$

Let

$$u = a_0 + a_1 y + a_2 y^2, \qquad (7.61)$$

where $(y')^2 = \sum_{j=0}^{4} c_j y^j = P_4(y)$.

Evidently, $u' = (a_1 + 2a_2 y)\sqrt{P_4(y)}$, $u^{''} = 2a_2 P_4 + (a_1 + 2a_2 y)\frac{P_4'}{2}$.

Substituting the expressions for $u^{'}$, $u^{''}$ into Eq. (7.60) we obtain the polynomial identity:

$$V(a_0+a_1y+a_2y^2) - \frac{A}{2}(a_0+a_1y+a_2y^2)^2 - E\left(2a_2P_4 + (a_1+2a_2y)\frac{p_4^{'}}{2}\right) \equiv 0.$$

Equating the coefficients of different powers of y^k, $0 \leq k \leq 4$ to be 0 we get the algebraic system of 5 nonlinear algebraic equations:

$$6a_2c_4E + A\frac{a_2^2}{2} = 0$$
$$(5a_2c_3 + 2a_1c_4)E + Aa_1a_2 = 0$$
$$(4a_2c_2 + 3a_1c_3/2)E + A(a_1^2 + 2a_0a_2)/2 - Va_2 = 0$$
$$-Va_1 + Aa_0a_1 + (a_1c_2 + 3a_2c_1)E = 0$$
$$-Va_0 + A\frac{a_0^2}{2} + (a_1c_1/2 + 2a_2c_0)E = 0.$$

Taking $c_0 = c_1 = c_4 = 0 = a_2 \Rightarrow c_2 = -\frac{V}{E}$, $c_3 = -\frac{Aa_1}{3E}$, $a_0 = \frac{2V}{A}$ we find for $VE < 0$, *i.e.* $c_2 > 0$ and $c_3 < 0$ that $u(\xi) = \frac{-3V}{A}(1 - tgh^2(\frac{\xi\sqrt{c_2}}{2})) + \frac{2V}{A}$, $1 - tgh^2\alpha = sech^2\alpha$.

Using the corresponding Tables of elliptic functions [30], [55] we can write down into explicit form several other solutions of quantum Zakharov-Kuznetsov equation. The details are left to the reader (see also [20]).

Remark 7.5. The elementary methods used for finding of solutions into explicit form are universal and powerful. Nevertheless, in some special cases (but very rare) they enable us to obtain the general solutions of the corresponding PDE. A good example in that direction is the following Liouville nonlinear hyperbolic PDE:

$$\frac{\partial^2 u}{\partial x \partial y} = ke^u, k > 0. \tag{7.62}$$

Let's differentiate Eq. (7.62) with respect to x and substitute $z = \frac{\partial u}{\partial x}$. Eliminating ke^u we get:

$$\frac{\partial^2 z}{\partial x \partial y} - \frac{1}{2}\frac{\partial}{\partial y}z^2 = 0, \tag{7.63}$$

i.e. Eq. (7.63) is equivalent to the Riccati ODE

$$\frac{\partial z}{\partial x} - \frac{1}{2}z^2 = f(x) \tag{7.64}$$

for some smooth function $f(x)$. Consider now the Cauchy problem

$$\frac{\varphi^{'''}(x)}{\varphi^{'}} - \frac{3}{2}\frac{(\varphi^{''})^2}{(\varphi^{'})^2} = f(x), \varphi(x_0) = a_0, \varphi^{'}(x_0) = b \neq 0, \varphi^{''}(x_0) = c. \tag{7.65}$$

Equation (7.65) possesses a smooth solution φ (locally) near x_0. Then for each $\psi(y) \in C^1$, $\varphi(x_0) + \psi(y_0) \neq 0$, $\psi'(y_0) \neq 0$, the function $z(x,y) = \frac{\varphi''}{\varphi'} - 2\frac{\varphi'(x)}{\varphi(x)+\psi(y)}$ satisfies Eq. (7.64). Check it. Therefore,

$$\frac{\partial u}{\partial x} = z(x,y) = \frac{\varphi''(x)}{\varphi'(x)} - 2\frac{\varphi'(x)}{\varphi(x)+\psi(y)}$$

implies that $u(x,y) = log\varphi' - 2log(\varphi + \psi) + C(y)$.

Taking $C(y) = log\psi'(y) + log\frac{2}{k}$, we get

$$\frac{k}{2}e^u = \frac{\varphi'\psi'}{(\varphi+\psi)^2}. \tag{7.66}$$

Conversely, for each φ, ψ smooth, $\varphi'(x_0)\psi'(y_0) > 0$, $\varphi(x_0) + \psi(y_0) \neq 0$ the function u defined by Eq. (7.66) satisfies Eq. (7.62) near (x_0, y_0)). (Check!) Therefore, the general solution of Eq. (7.62) is given by Eq. (7.66), φ, ψ being arbitrary, $\varphi'\psi' > 0$, $\varphi + \psi \neq 0$.

7.3 Traveling waves for some generalized Boussinesq type equations

This section deals with the $(N + 1)$ dimensional dispersive Boussinesq equation (including the shallow water waves equation obtained for $n = 3$, $N = 2$):

$$u_{tt} = u_{xx} + \lambda(u^n)_{xx} + \mu(u^m)_{xxxx} + \sum_{j=1}^{N-1} u_{y_j y_j}, \tag{7.67}$$

where $N \geq 2$, $\lambda = $ const, $\mu = $ const, $\lambda\mu \neq 0$, $n \geq 2$, $m \geq 1$, $m, n \in \mathbf{N}$, $u = u(t, x, y_1, \ldots, y_{N-1})$, $(t, x) \in \mathbf{R}^2$ and the $B(m, n)$ equation

$$u_{tt} = u_{xx} + \lambda(u^n)_{xx} + \mu(u^m)_{xxxx},$$

$u = u(t, x)$, $(t, x) \in \mathbf{R}^2$, $m > 1$, $n > 1$, $m, n \in \mathbf{N}$, $\lambda\mu \neq 0$, $\lambda, \mu = $ const.
$$\tag{7.68}$$

The classical Boussinesq equation coincides with $B(1, 2)$. We put $u = u(t, x, y)$ for the solution of Eq. (7.67) and we shall find the traveling wave solutions of Eq. (7.67), Eq. (7.68) into integral form. Moreover, their regularity properties and their singularities (if they exist) will be studied. In some special cases we obtain solutions written in the form of Jacobi elliptic functions $sn(\xi)$, $cn(\xi)$, $dn(\xi)$ or in the form of Legendre elliptic functions of I, II and III kinds. We propose below several interesting papers on the Boussinesq equation published in [68], [85], [137], [140], [149].

To complete the Introduction we point out that following [57] we shall look for the solutions of Eq. (7.67) of the form:

$$u = \varphi(\xi), \text{ where } \xi = \tau \left(x + \sum_1^{N-1} y_j - ct \right), \ \tau = \text{const} \neq 0, \ c = \text{const} \neq 0$$

(7.69)

and for the equation $B(m, n)$ we put

$$u = \varphi(\xi), \text{ where } \xi = x - ct, \ c = \text{const} \neq 0.$$

(7.70)

7.3.1 Construction of traveling wave solutions to Boussinesq type PDE

1. Substituting Eq. (7.69) in Eq. (7.67) we get the ODE

$$\varphi''(c^2 - N) - \lambda(\varphi^n)'' - \mu(\varphi^m)^{\text{IV}}\tau^2 = 0.$$

(7.71)

Integrating Eq. (7.71)) twice with respect to ξ and taking one of the integral constants to be equal to 0 we obtain:

$$\tau^2\mu(\varphi^m)'' + \lambda(\varphi^n) + \varphi(N - c^2) = \frac{A}{2} = \text{const.}$$

(7.72)

We multiply Eq. (7.72) by $(\varphi^m)'$ and integrate Eq. (7.72) again with respect to ξ. Thus,

$$\frac{m}{m+1}(N-c^2)\varphi^{m+1} + \frac{m}{m+n}\lambda\varphi^{m+n} + \frac{1}{2}\mu\tau^2(\varphi^m)'^2 = \frac{A}{2}\varphi^m + \frac{B}{2}, \ B = \text{const},$$

(7.73)

i.e.

$$(\varphi^m)'^2 = c_0\varphi^{m+n} + d_0\varphi^{m+1} + A_1\varphi^m + B_1 \equiv Q_{m+n}(\varphi),$$

(7.74)

where $c_0 = -\dfrac{2m\lambda}{\mu\tau^2(m+n)} \neq 0, \ d_0 = \dfrac{2m(c^2 - N)}{(m+1)\mu\tau^2}, \ A_1 = \dfrac{A}{\mu\tau^2}, \ B_1 = \dfrac{B}{\mu\tau^2}.$
Certainly, $Q_{m+n}(\varphi) \geq 0$ for each $\varphi(\xi)$, ξ belonging to the definition interval of the solution φ.

Remark 7.6. Let $m = 1$. Then

$$\varphi'^2 = c_0\varphi^{n+1} + d_0\varphi^2 + A_1\varphi^m + B_1,$$

(7.75)

while for $m \geq 2$ we have

$$m^2\varphi^{2m-2}\varphi'^2 = Q_{m+n}(\varphi).$$

(7.76)

The latter equation was studied in detail in Chapter II of [110].

Consider the equation Eq. (7.75). It can be solved easily for $n = 1$ via the Euler substitutions. Elliptic functions appear for $n = 2$ and $n = 3$.

Example 1. Assume that $A = 0$ and $\tau = 1$ for $n = 3$. Then we get:

$$\varphi'^2 = c_0\varphi^4 + d_0\varphi^2 + B_1. \tag{7.77}$$

On the other hand, as we know in the paper [57] (see also [44]) a Table containing the expressions of the solutions of the ODE: $z'^2 = c_1 + c_2 z^2 + \frac{c_3}{2} z^4$ via the Jacobi elliptic functions is given. According to the case No 1 from Table 1 [57]

$$\lambda = -2r^2 \frac{N - c^2}{r^2 + 1}, \quad \mu = \frac{N - c^2}{r^2 + 1}, \quad B = \mu, \quad 0 < r < 1$$

imply that Eq. (7.77) possesses the periodic solutions $\varphi(\xi) = \mathrm{sn}(\xi, r)$ and $\varphi = \mathrm{cd}(\xi, r) = \frac{\mathrm{cn}(\xi, r)}{\mathrm{dn}(\xi, r)}$.

Another 19 examples illustrating Eq. (7.77) can be deduced from the above mentioned Table 1.

2. To simplify the things we shall assume that $\lambda = 1$, $\mu = -1$ in Eq. (7.68), i.e. $B(m, n)$ has the form:

$$u_{tt} - u_{xx} + (u^m)_{xxxx} - (u^n)_{xx} = 0, \tag{7.78}$$

m, n — being arbitrary integers, $m > 1$, $n > 1$. The case $m = n$ was studied via polynomials of elliptic function in [68]. Having in mind Eq. (7.70) we obtain that $\varphi(\xi)$ satisfies

$$\varphi^{m'2} = c_0\varphi^{m+n} - d_0\varphi^{m+1} + A\varphi^m + B \equiv \tilde{Q}_{m+n}(\varphi) \tag{7.79}$$

$A, B = \text{const}$, $c_0 = \dfrac{3m}{m+n}$, $d_0 = \dfrac{2m}{m+1}(c^2 - 1)$ and the polynomial $\tilde{Q}_{m+n}(\varphi) \geq 0$.

We shall consider two cases, namely

Case I: $d_0 > 0 \Leftrightarrow c^2 > 1$, $A = B = 0$, m — even, $n \geq 3$ — odd integers.

Case II: $B = 0$, $A < 0$, $n \geq 2$, even, $m \geq 3$ — odd integers.

3. We shall concentrate at first to the Case I. Then $\tilde{Q}_{m+n}(\varphi) = \varphi^{m+1}(c_0\varphi^{n-1} - d_0)$, $\mathrm{sgn}\,\varphi^{m+1} = \mathrm{sgn}\,\varphi$ and the only real roots of the algebraic equation $\tilde{Q}_{m+n}(\varphi) = 0$ are: $\varphi_1 = 0$ (multiple one), $\varphi_2 = \left(\dfrac{d_0}{c_0}\right)^{\frac{1}{n-1}} > 0$, $\varphi_3 = -\varphi_2$ (both are simple roots, i.e. $\tilde{Q}'_{m+n}(\varphi_i) \neq 0$; $i = 2, 3$).

If we are looking for a bounded solution of Eq. (7.79) we should confine ourselves with $\varphi \in (\varphi_3, 0) \cup (0, \varphi_2)$. Having in mind the graph of the polynomial $\tilde{Q}_{m+n}(\varphi)$ we conclude that solutions $\varphi(\xi)$ of Eq. (7.79) with $\varphi(0) < \varphi_3$, respectively $0 < \varphi(0) < \varphi_2$ do not exist, while the solution φ with $\varphi(0) > \varphi_2$ is unbounded. Therefore, bounded solution exists in the Case I iff $\varphi_3 < \varphi < 0$. Consequently, we have to study the degenerate ODE

$$\varphi' = \frac{\sqrt{\tilde{Q}_{m+n}(\varphi)}}{m\varphi^{m-1}} = \frac{\sqrt{\tilde{Q}_{m+n}(\varphi)}}{-m|\varphi|^{m-1}}, \quad 0 > \varphi > \varphi_3 \tag{7.80}$$

with initial condition $\varphi(0) = \varphi_3 < 0$; $m - 1$ is odd. Evidently, $\varphi(\xi)$ is monotonically decreasing function in its definition domain. Moreover,

$$\xi = m \int_{\varphi_3}^{\varphi(\xi)} \frac{\lambda^{m-1} d\lambda}{\sqrt{\tilde{Q}_{m+n}(\lambda)}} = F(\varphi) < 0 \quad \text{for} \quad \varphi_3 < \varphi < 0. \tag{7.81}$$

Certainly, $F \in C^\infty((\varphi_3, 0))$ and

$$F(\varphi) = -m \int_{\varphi_3}^{\varphi} \frac{|\lambda|^{\frac{m-3}{2}}}{\sqrt{d_0 - c_0|\lambda|^{n-1}}}, \quad F'(\varphi) < 0. \tag{7.82}$$

Having in mind that $m \geq 2 \Rightarrow \frac{m-3}{2} \geq -1/2$ the integral Eq. (7.82) is convergent at $\varphi = -0 \Rightarrow$

$$-\infty < F(-0) < 0, \quad \lim_{\varphi \to -0} F'(\varphi) = \begin{cases} -\infty, & m = 2 \\ -0, & m \geq 4 \end{cases} \tag{7.83}$$

Put $F(-0) = T < 0 \Rightarrow F(\varphi) = T + \int_{-0}^{\varphi} \frac{-m|\lambda|^{\frac{m-3}{2}} d\lambda}{\sqrt{d_0 - c_0|\lambda|^{n-1}}}$.

Thus,

$$0 > F(\varphi) \sim T + \frac{m}{\sqrt{d_0}} \cdot \frac{2}{m-1} |\varphi|^{\frac{m-1}{2}} \quad \text{for} \quad \varphi \sim 0, \quad \varphi < 0. \tag{7.84}$$

As it concerns the behavior of $F(\varphi)$ for $\varphi \to \varphi_3 + 0$ we write $n - 1 = 2k$, $k \in \mathbf{N}$ and therefore

$$-\lambda^{n-1} + \frac{d_0}{c_0} = (-\lambda + \varphi_2)(\lambda - \varphi_3)Q_{n-3}(\lambda),$$

Q_{n-3} being nonvanishing polynomial of even order $n - 3$, i.e. $Q_{n-3}(\lambda) \geq \bar{c} > 0$ for some constant $\bar{c}$. Thus, $\lambda \sim \varphi_3 + 0 \Rightarrow |\lambda|^{\frac{m-3}{2}} \sim |\varphi_3|^{\frac{m-3}{2}}$ and $\sqrt{d_0 - c_0\lambda^{n-1}} \sim \sqrt{c_0}\sqrt{2\varphi_2}\sqrt{\lambda - \varphi_3}\bar{d_0}$, where $\bar{d_0} = \text{const} > 0$.
Therefore,

$$F(\varphi) \sim \text{const} \sqrt{\varphi - \varphi_3} \quad \text{for} \quad \varphi \to \varphi_3 + 0 \quad \text{and const} < 0. \tag{7.85}$$

Combining Eq. (7.80), Eq. (7.82), Eq. (7.83), Eq. (7.84) and Eq. (7.85) we obtain the graph of the inverse function $\varphi = F^{-1}(\xi) \in C^\infty(T, 0)$ of Eq. (7.81) (Figs. 7.1, 7.2):

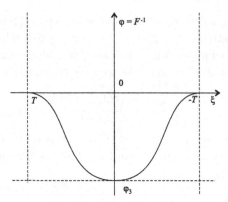

Fig. 7.1. Case I, $m = 2$.

Having constructed $\varphi(\xi)$ for $T < \xi < 0$ we make an even continuation of $\varphi(\xi)$ in the interval $(O, -T)$, i.e. $\varphi(\xi) = \varphi(-\xi)$, which satisfies Eq. (7.79)) in $(0, -T)$, $\varphi(0) = \varphi_3$ and $\varphi_3 < \varphi(-\xi) < 0$. We prolong φ from $[T, -T]$ to $\mathbf{R}^1$ as a periodic function with period $-2T$. We remind that compacton is a traveling wave having compact support.

Theorem 7.1. [108]

Consider the $B(m, n)$ equation Eq. (7.78) in the Case I. Then there exists a smooth compacton solution (smooth periodic solution) if and only if $m = 2$. In the case $m \geq 4$, m — even, the solution $\varphi(\xi)$ is of the type compacton-cuspon with cusp type singularities for $|\xi| = |T|$ (periodic function with cusp singularities).

Remark 7.7. We claim that in Case I $\varphi(\xi)$ satisfies in classical sense in $\mathbf{R}^1$ the ODE $(c^2 - 1)\varphi + (\varphi^m)'' - \varphi^n = 0$, $m \geq 4$, while the ODE $(c^2 - 1)\varphi'' + (\varphi^m)^{\mathrm{IV}} - (\varphi^n)'' = 0$ corresponding to the PDE Eq. (7.78) is satisfied by φ in Schwartz distributional sense (i.e. in $D'(\mathbf{R}^1)$). To fix the ideas we assume that φ is a compacton. Having in mind that $\varphi \in C^2((+T, -T))$ we have near to the singular point $\xi = T, \xi > T$: $\varphi \sim -\dfrac{(\xi - T)^{2/m-1}}{f_m^{2/m-1}} \Rightarrow (c^2 - 1)\varphi + (\varphi^m)'' - \varphi^n \sim A(\xi - T)^{\frac{2}{m-1}}, \xi \to T+0, A = \mathrm{const} \neq 0$, while $\varphi = 0$ for $\xi < T$, etc. It is well-known that for each $a \in \mathbf{R}$ one can define a distribution x_+^a, supp $x_+^a \in \{x \geq 0\}$ and such that the generalized derivative $\dfrac{d}{dx} x_+^a = a x_+^{a-1}$ for $a \notin \{0, -1, -2, -3, \dots\}$, while $a = -k \in \{0, -1, -2, \dots\} \Rightarrow$

$$\frac{d}{dx}x_+^{-k} = -kx_+^{-k-1} + (-1)^k\frac{\delta^{(k)}(x)}{k!}, \ \delta \text{ being the Dirac } \delta \text{ function. The}$$

distribution $\varphi_+ = \begin{cases} A(\xi - T)^{\frac{-2}{m-1}}, \xi > T, \xi \approx T \\ 0, \qquad\qquad \xi \le T \end{cases}$ is, of course, differentiable

in D' and $\delta^{(j)}(\xi - T)$ does not appear after the differentiation. Therefore, $(c^2 - 1)\varphi'' + (\varphi^m)^{IV} - (\varphi^n)'' = 0$ in $D'(\mathbf{R}^1)$.

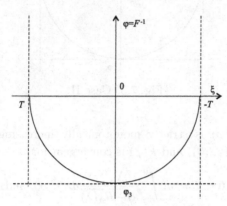

Fig. 7.2. Case I, $m \ge 4$.

4. The Case II is simple (see Fig. 7.3). In fact,

$$(\varphi^m)'^2 = \widetilde{\widetilde{Q}}_{m+n}(\varphi) = \varphi^m(c_0\varphi^n - d_0\varphi + A) = \varphi^m R_n(\varphi), \qquad (7.86)$$

$A < 0, c_0 > 0, d_0 > 0, n$ — even and m is odd integer.

Evidently, $\varphi_1 = 0$ is a multiple root of $\widetilde{\widetilde{Q}}_{m+n}(\varphi) = 0$. From geometrical reasons it is clear that $R_n(\varphi) = 0$ possesses only two real simple roots $\varphi_3 < 0$, $\varphi_2 > 0$. Moreover, $R_n(\varphi) > 0$ for $\varphi < \varphi_3$ and for $\varphi > \varphi_2$, while $\varphi_3 < \varphi < \varphi_2 \Rightarrow R_n(\varphi) < 0$. If we are looking for bounded solutions we shall consider the following Cauchy problem

$$(\varphi^m)'^2 = \varphi^m R_n(\varphi), \quad \varphi_3 < \varphi < 0, \quad \varphi(0) = \varphi_3 < 0, \quad m \text{ — odd.} \quad (7.87)$$

We shall concentrate on one branch of the $\sqrt{\varphi^m R_n(\varphi)}$, namely

$$\varphi' = \frac{\sqrt{-A + d_0\varphi - c_0\varphi^n}}{m|\varphi|^{\frac{m-2}{2}}}, \quad \varphi(0) = \varphi_3, \ 0 > \varphi > \varphi_3. \qquad (7.88)$$

Obviously, $\varphi(\xi)$ is strictly monotonically increasing,

$$\xi = m\int_{\varphi_3}^{\varphi} \frac{|\lambda|^{\frac{m-2}{2}}d\lambda}{\sqrt{-A + d_0\lambda - c_0\lambda^n}} = F(\varphi) > 0. \qquad (7.89)$$

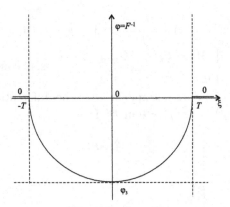

Fig. 7.3. Case II.

Therefore, $F(\varphi)$ is strictly monotonically increasing in the interval $(\varphi_3, 0)$, $F \in C^{\infty}((\varphi_3, 0))$ and $F(\varphi)$ is convergent for $\varphi \to -0$, i.e.

$$0 < T = m \int_{\varphi_3}^{0} \frac{|\lambda|^{\frac{m-2}{2}}}{\sqrt{-R_n(\lambda)}} d\lambda < \infty \quad \text{and}$$

$$F(\varphi) = T + m \int_{0}^{\varphi} \frac{|\lambda|^{\frac{m-2}{2}} d\lambda}{\sqrt{-R_n(\lambda)}} \sim T - \frac{2}{\sqrt{|A|}} |\varphi|^{\frac{m}{2}} \quad \text{for} \quad \varphi \to -0.$$

We have that for $\varphi_3 < \lambda < \varphi_2$: $0 < -A + d_0\lambda - c_0\lambda^n = -c_0(\lambda - \varphi_3)(\lambda - \varphi_2)M_{n-2}(\lambda)$ and the polynomial of even order $n - 2$: $M_{n-2}(\lambda) > 0$ in the interval $[\varphi_3, \varphi_2]$. Thus, $\varphi \sim \varphi_3 + 0 \Rightarrow$

$$0 < F(\varphi) \sim B\sqrt{\varphi - \varphi_3} \quad \text{for} \quad \varphi \to \varphi_3 + 0 \quad \text{and} \quad B = \text{const} > 0. \quad (7.90)$$

Combining Eq. (7.88), Eq. (7.89), Eq. (7.90) we obtain $\varphi = F^{-1}(\xi)$:

Theorem 7.2. [108]

In the Case II there exists a compacton — cuspon solution of the $B(m, n)$ equation Eq. (7.78), $\varphi(-\xi) = \varphi(\xi)$ as well as periodic cusp wave solution. The cusp singularity appears for $|\xi| = T$.

Compacton-cuspon is a compacton developing cusp type singularities.

7.4   Interaction of two solitons and rogue waves

1. At first we shall briefly discuss the problem of the interaction of solitary waves satisfying the 1D version of Eq. (7.22), where $f(u) = u^2$, $\alpha_0 = 3$, $\beta_1 = 1.5$ and $\beta_2 = 0.5$, $m = 1$.

From Eq. (7.29) we know that Eq. (7.22) possesses the traveling wave solution (soliton):

$$\tilde{\varphi}(x,t;x_0,c) = \frac{3}{2}\frac{c^2-1}{\alpha_0}sech^2\left(\frac{x-x_0-ct}{2}\sqrt{\frac{c^2-1}{\beta_1 c^2-\beta_2}}\right),$$

$|c| > max(1,\sqrt{\frac{\beta_2}{\beta_1}})$ or $|c| < min(1,\sqrt{\frac{\beta_2}{\beta_1}})$. The maximum of $\tilde{\varphi}$ is attained at the line $x-x_0-ct=0$.

Consider now the Cauchy problem for Eq. (7.22) with initial data

$$u(x,0) = \tilde{\varphi}(x,0;x_0^1,c_1) + \tilde{\varphi}(x,0;x_0^2,c_2), \qquad (7.91)$$
$$u_t(x,0) = \frac{\partial\tilde{\varphi}}{\partial t}(x,0;x_0^1,c_1) + \frac{\partial\tilde{\varphi}}{\partial t}(x,0;x_0^2,c_2), x \in \mathbf{R}^1.$$

As Eq. (7.22) is nonlinear PDE, $u(x,t) \not\equiv \tilde{\varphi}(x,t;x_0^1,c_1) + \tilde{\varphi}(x,t;x_0^2,c_2)$ in the general case.

By applying a new conservative finite difference scheme for Eq. (7.22) the following results were found in [80]

Case a). $\beta_1 = 1.5$, $\beta_2 = 0.5$, $\alpha_0 = 3$, $x_0^1 = -40$, $x_0^2 = 50$, $c_1 = 2$, $c_2 = -1.5$, $0 \le t \le 90$.

Thus, $\tilde{\varphi}(x,0;x_0^1,c_1)$ and $\tilde{\varphi}(x,0;x_0^2,c_2)$ are two solitary waves moving in opposite directions. They collide and it seems that due to their interaction a new wave appears. It is weaker than the two initial waves, i.e. the corresponding amplitudes are smaller. Both initial waves keep traveling preserving their shapes after the collision for increasing time. This way, two mathematical problems appear: To prove the appearance of a new born wave and to verify that it is a dying one (see Fig. 7.4).

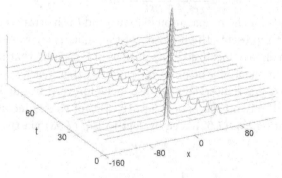

Fig. 7.4

Case b). $\beta_1 = 1.5$, $\beta_2 = 0.5$, $\alpha_0 = 3$, $x_0^1 = -40$, $x_0^2 = 50$, $c_1 = -c_2 = 2.2$.

Then the solution u blows up after the collision increasing the absolute value of the amplitude. The blow up time $t^* \approx 27$ (see Fig. 7.5).

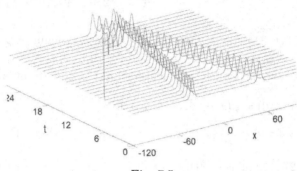

Fig. 7.5

As one can guess the different velocities give rise to different effects. Interaction of 3 traveling waves in $\mathbf{R}_t^1 \times \mathbf{R}_x^2$ and new born wave carrying out weaker singularities was studied via paradifferential operators by J. M. Bony about 1985. Elementary proof in some special cases can be found in [106].

2. Consider now the class of the so-called "rogue waves" [100], "freak waves" or "killer waves". They are giant single waves that appear in the ocean "from nowhere" and disappear "without a trace". It is also known that rogue waves can be generated by optical systems creating another possibility for the generation of highly energetic optical pulses. The above mentioned waves are described by the nonlinear Schrödinger equation

$$i\frac{\partial \psi}{\partial x} + \frac{1}{2}\frac{\partial^2 \psi}{\partial t^2} + \psi|\psi|^2 = 0, \qquad (7.92)$$

where x stands for the propagation distance and t is its transversal variable. In optics $|\psi|^2$ represents the intensity of the light. As it was shown in [11] there exists a class of localized in x solutions of Eq. (7.92) that increase their amplitudes exponentially or polynomially in x, then reach their maximum value and decay symmetrically to disappear forever. In our investigation below we shall rely on the papers [9], [10], [11] of N. Akhmediev et al.

The equation Eq. (7.92) is equivalent to Eq. (7.93) via the change

$$\psi = ue^{ix},$$

where

$$iu_x + \frac{1}{2}u_{tt} + u(|u|^2 - 1) = 0. \qquad (7.93)$$

If u satisfies Eq. (7.93) then $\tilde{u} = \pm iu$ is a solution of the same equation. One can easily see that if ψ satisfies Eq. (7.92) then $\tilde{\psi} = q\psi(qt, q^2x)$ is a solution of Eq. (7.92) with q being a real parameter.

Put $u = v + iw$. Evidently, $v = \pm 1$, $w = 0$ is a solution of Eq. (7.93) and the real valued functions v, w verify the system

$$v_x + \frac{1}{2}w_{tt} - w + w(v^2 + w^2) = 0 \tag{7.94}$$

$$-w_x + \frac{1}{2}v_{tt} - v + v(v^2 + w^2) = 0$$

We are looking for a solution (v, w) of Eq. (7.94) having the form

$$w = (v - 1)\eta(x), \tag{7.95}$$

i.e. after eliminating v_{tt} from Eq. (7.94) we get:

$$v_x - (v - 1)\eta + \eta(v - 1)(v^2 + (v - 1)^2\eta^2) + v_x\eta^2 + (v - 1)\eta\eta_x + \eta v - \tag{7.96}$$

$$\eta v(v^2 + (v - 1)^2\eta^2) = 0.$$

Easy computations transform Eq. (7.96) into

$$v_x - \eta v^2 + v\eta\frac{(\eta_x + 2\eta^2)}{1 + \eta^2} + \eta\frac{(1 - \eta_x - \eta^2)}{1 + \eta^2} = 0. \tag{7.97}$$

Assuming that η satisfies the Riccati equation

$$\eta_x + \eta^2 = 1, \tag{7.98}$$

i.e. $\eta = \pm 1$, $\eta = tgh(x + C)$, $\eta = ctgh(x + C)$, $C = const$ we obtain from Eq. (7.97) the Bernoulli equation

$$v_x - \eta v^2 + \frac{v\eta(\eta_x + 2\eta^2)}{1 + \eta^2} = 0 \tag{7.99}$$

In fact, putting $\eta = tgx$ in Eq. (7.99) we come to

$$v_x - tghxv^2 + vtghx = 0. \tag{7.100}$$

That standard change $z = \frac{1}{v}$ reduces Eq. (7.100) to a first order linear ODE with respect to the variable x, t being a parameter. Thus, $z = \frac{1}{d(t)}chx + 1$, i.e.

$$v = \frac{d(t)}{chx + d(t)} \Rightarrow v - 1 = -\frac{chx}{chx + d(t)}. \tag{7.101}$$

To find the unknown $d(t)$ we substitute Eq. (7.101) into the equation

$$v_x + \frac{1}{2}v_{tt}\eta - (v - 1)\eta + (v - 1)\eta(v^2 + (v - 1)^2\eta^2) = 0, \eta = tghx. \tag{7.102}$$

Standard computations with hyperbolic functions reduce Eq. (7.102) to the equation Eq. (7.103):

$$d''(t) - \frac{2(d')^2(t)}{chx + d(t)} + 2\frac{1 - d^2 + dchx}{chx + d(t)} = 0. \tag{7.103}$$

Differentiating Eq. (7.103) with respect to x we come to the ODE with separate variables

$$(d')^2 = 1 - 2d^2, \tag{7.104}$$

i.e. $d = \pm\frac{1}{\sqrt{2}}sin(t\sqrt{2})$ or $d = \pm\frac{1}{\sqrt{2}}cos(t\sqrt{2})$.

Therefore, if $d = +\frac{1}{\sqrt{2}}cos(t\sqrt{2})$, $\eta = tghx$ then

$$v(t,x) = \frac{cos(t\sqrt{2})}{\sqrt{2}chx + cos(t\sqrt{2})} \tag{7.105}$$

According to: $\psi = ue^{ix}$, $u = v + iw$, $w = (v - 1)\eta(x)$ and Eq. (7.105)

$$\psi = \frac{cos\sqrt{2}t - \sqrt{2}ishx}{\sqrt{2}chx + cos(\sqrt{2}t)}e^{ix}. \tag{7.106}$$

Certainly, $cos(\sqrt{2}t) + \sqrt{2}chx \geq \sqrt{2} - 1 > 0$.

The following interesting result is left to the reader as an exercise (see [9]).

Exercise 11. Check that the function

$$\psi = \frac{(1 - 4a)ch(\beta x) + \sqrt{2a}cos(pt) + i\beta sh(\beta x)}{\sqrt{2a}cos(pt) - ch(\beta x)}e^{ix},$$

where $\beta = \sqrt{8a(1 - 2a)}$, $p = 2\sqrt{1 - 2a}$, $0 < a < 1/2$ satisfies Eq. (7.92). Find $lim_{a \to 1/2}\psi(x, t)$.

Hint. Let $a \to 1/2$, t, x being fixed. Then $ch(\beta x) \sim 1 + \frac{\beta^2 x^2}{2}$, $\beta \to 0$; $cos(pt) \sim 1 - \frac{p^2 t^2}{2}$, $p \to 0$, $\beta sh(\beta x) \sim \beta^2 x$, $\beta \to 0$. Having in mind that $1 - 2a = (1 - \sqrt{2a})(1 + \sqrt{2a})$, $1 - 4a + \sqrt{2a} = -2(\sqrt{2a} - 1)(\sqrt{2a} + 1/2)$ after some elementary algebra one gets that

$$lim_{a \to 1/2}\psi(t, x) = \frac{-3 + 4(t^2 + x^2) - i8x}{(1 + 4(t^2 + x^2))}e^{ix} = \left(1 - 4\frac{1 + 2ix}{1 + 4(t^2 + x^2)}\right)e^{ix}. \tag{7.107}$$

Evidently, $\psi \in C^\infty(\mathbf{R}^2)$ is a rational solution of Eq. (7.92). Moreover,

$$|\psi|^2 - 1 = 8\frac{1 + 4(x^2 - t^2)}{(1 + 4(x^2 + t^2))^2} = z. \tag{7.108}$$

The cylindrical change $z = z$, $x = \rho cos\varphi$, $t = \rho sin\varphi$, $0 \leq \varphi \leq 2\pi$ gives us that

$$z(x,t) = 8\frac{1 + 4\rho^2 cos2\varphi}{(1 + 4\rho^2)^2},$$

i.e. $\rho \to \infty \Rightarrow z \to 0$ ($\Rightarrow |\psi| \to 1$ for $\rho \to \infty$), $8\frac{1-4\rho^2}{(1+4\rho^2)^2} \leq z \leq \frac{8}{1+4\rho^2}$.

Evidently, $z = \frac{8}{1+4\rho^2} \iff \varphi = 0, \pi$, $z = 8\frac{1-4\rho^2}{(1+4\rho^2)^2} \iff \varphi = \frac{\pi}{2}, \frac{3}{2}\pi$. The maximum value of z, $z = 8$ is reached at $\rho = 0$, $\varphi = 0, \pi$ and its minimum $z = -1$ is reached at $\rho = \frac{\sqrt{3}}{2}$, $\varphi = \frac{\pi}{2}, \frac{3}{2}\pi$. The function z changes its sign at the hyperbola $4\rho^2 = -\frac{1}{cos2\varphi}$.

Conclusion: $max_{\mathbf{R}^2}|\psi| = 3$, $min_{\mathbf{R}^2}|\psi| = 0$, $|\psi| \to_{\rho \to \infty} 1$, $|\psi| = 1 \iff 1 + 4\rho^2 cos2\varphi = 0 \iff x^2 - t^2 = -1/4$ (see Fig. 7.6).

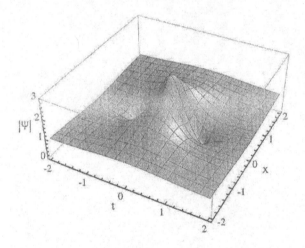

Fig. 7.6.

Chapter 8

Regularity properties of several hyperbolic equations and systems

8.1 Introduction

This chapter deals with the regularity properties (including propagation and interaction of nonlinear waves) of the solutions of the Cauchy problem to 2D semilinear wave equation with the removable singularities of the solutions of fully nonlinear hyperbolic systems arising in the mechanics of compressible fluids with constant entropy, and with the regularizing properties of the multidimensional wave equation with dissipative term. In the first two cases discussed in Section 2 and Section 3 the machinery of the pseudodifferential, respectively paradifferential operators is applied ([24], [67], [127]). More precisely, "radially smooth" initial data having singularities on a "massive" set of angles in the plane, including the Cantor continuum set, yield singularities propagating as in the linear case [18]. There is a big difference between the 2D case and the multidimensional case (3D) when the interaction of several (for example four) characteristic hyper-planes could produce singularities on a dense subset of the compliment of the light cone of the future located over the origin (see [18] for more details). A result of Bony [23], [25] for the triple interaction of progressing linear waves in the 2D case is commented too as then new effect appears: new born wave propagating along the cone of the future with vertex at the origin. In Section 3 we assume that the first variation of the nonlinear system under consideration is linear, symmetric and positive one in the sense of Friedrichs. A microlocal version of the Moser's condition on the existence of global solutions on the torus of the same system [98] enables us to prove the nonexistence of isolated singularities at each characteristic point of the main symbol of the first variation. For symmetric quasilinear hyperbolic systems we study the propagation of regularity. As usual, the

155

strength of the singularities is measured both in Sobolev H^s and microlocalized Sobolev spaces H^s_{mcl}. An example from fluid mechanics illustrates our results [12].

In Sections 4–5 we consider the multidimensional $(n > 3)$ wave equation with the dissipative term $|u_t|^{h-1}u_t$, $h > 1$. Then the piecewise smooth solution $u \in C^0$ of that equation, having finite jumps of ∇u for $t < 0$ becomes C^1 smooth for $t > 0$. On the other hand, in the case $n = 3$ a solution u with logarithmic singularities for $t < 0$ develops C^1 regularity for $t > 0$. The Chapter is illustrated by several examples and figures. The proofs are not complicated but still require some efforts for understanding.

8.2 Regularity results on the 2D semilinear wave equation having radially smooth Cauchy data

We consider solutions to the 2D wave equation and to the corresponding Cauchy problem:

$$\Box u = u_{x_3 x_3} - u_{x_1 x_1} - u_{x_2 x_2} = f(x, u, \nabla_x u) \tag{8.1}$$
$$\text{on } \Omega, f(x, u, p) \in C^\infty,$$

$$u|_{x_3=0} = u_0(x_1, x_2), u_{x_3}|_{x_3=0} = u_1(x_1, x_2). \tag{8.2}$$

Denote by M the radial vector field $\sum_{i=1}^3 x_i \frac{\partial}{\partial x_i}$ and by $\tilde{M} = x_1 \frac{\partial}{\partial x_1} + x_2 \frac{\partial}{\partial x_2}$ the spatial radial vector field. The domain $\Omega \subset \mathbf{R}^3$ contains 0, $\Omega = \Omega^+ \cup \tilde{\Omega} \cup \Omega^-$, $\Omega^+ = \Omega \cap \{x_3 > 0\}$, $\Omega^- = \Omega \cap \{x_3 < 0\}$, $\tilde{\Omega} = \Omega \cap \{x_3 = 0\}$. The symbol $\tilde{\Omega}^+$ stands for an open set in $\mathbf{R}^2$, such that $\bar{\tilde{\Omega}}^+$ is a compact in $\tilde{\Omega}$. As usual, $K = \{x_3^2 = x_1^2 + x_2^2\}$ is a light cone. Evidently, M is tangential to the characteristic cone K and to the characteristic hyperplane of $\Box : \alpha : x_3 = <w, x'>$, $w \in \mathbf{R}^2$, $|w| = 1$, $x' = (x_1, x_2)$.

If we consider Eq. (8.1) in $T^*(\Omega)$, then its principal symbol is $p_2(x, \xi) = \xi_3^2 - |\xi'|^2$. The zero bicharacteristic curves of p_2 are given by

$$\gamma \left| \begin{array}{l} \dot{x} = \frac{\partial p_2}{\partial \xi}, \quad x(0) = x_0, |t| < \varepsilon \\[2mm] \dot{\xi} = -\frac{\partial p_2}{\partial x}, \xi(0) = \xi_0, \xi_{30}^2 = |\xi_0'|^2 > 0 \Rightarrow p_2(x_0, \xi_0) = 0 \Rightarrow p_2|_\gamma = 0. \end{array} \right.$$

Thus, $x_3 = +2\xi_{30}t + x_{30}$, $x' = -2\xi_0' t + x_0'$, $\xi_{30}^2 = |\xi_0'|^2$.

Moreover, $|x' - x_0'|^2 = (x_3 - x_{30})^2$, i.e. the projection of the bicharacteristic γ on $\mathbf{R}_x^3$ is a generatrix of the characteristic cone with vertex at x_0. It is well known (see [67], [127]) that if $\gamma \cap WF(\Box u) = \emptyset$ then either $\gamma \subset WF(u)$ or $\gamma \cap FW(u) = \emptyset$. We propose here a definition of $H_{mcl}(\rho^0)$,

$\rho^0 = (x^0, \xi^0)$, $\xi^0 \neq 0$. There exists cut of function $\varphi(x)$, $\varphi \equiv 1$ near x^0 and such that $\int_\Gamma (1 + |\xi|^2)^s |\widehat{\varphi u}(\xi)|^2 d\xi < \infty$, where Γ is a closed cone in $\mathbf{R}^3_\xi$ with vertex at the origin and axes ξ^0. The symbol $\widehat{}$ stands for the Fourier transform of φu. Sometimes instead of $u \in H^s_{mcl}(\rho^0)$ it is written $\rho^0 \notin WF_s(u)$; $WF(u) = \overline{\cup_s WF_s(u)}$. $WF(u)$ is called wave front set of u.

If the surface S is given by $\{x_1 = 0\}$ then the tangential space to S is generated by the vector fields $x_1 \frac{\partial}{\partial x_1}, \frac{\partial}{\partial x_2}, \dots, \frac{\partial}{\partial x_n}$.

We shall apply several auxiliary results from the (microlocal) analysis further on. To make easier the acceptance of the proof of our main result in this section we shall formulate them.

Lemma 8.1. *(Schauder)* [117]

Assume that $f(x, u) \in C^\infty$ *and* $u \in H^s_{loc}(\mathbf{R}^n)$, $s > n/2$. *Then* $f(x, u(x)) \in H^s_{loc}(\mathbf{R}^n)$.

The result can be generalized in the following form: $u \in H^s_{loc} \cap L^\infty_{loc}$, $s \geq 0$ implies that $f(x, u(x)) \in H^s_{loc}$ if $f(x, 0) = 0$ (see [117]).

As it is proved in [117] if $s > 2.5$, $f(0, 0, 0) = 0$ the equation Eq. (8.1) is locally solvable in Ω, while the Cauchy problem is uniquely solvable for some $T \in [0, 1)$ in the space $C([0, T); H^s(\tilde{\Omega}^+)) \cap C^1([0, T); H^{s-1}(\tilde{\Omega}^+))$ for Cauchy data $u_0 \in H^s(\tilde{\Omega})$, $u_1 \in H^{s-1}(\tilde{\Omega})$.

Proposition 8.1.

Suppose that $s > 2.5$ *and* $u \in H^s(\Omega)$ *satisfies Eq. (8.1). If* $M^j u \in H^s(\Omega^-)$ *for all* $j \in \mathbf{N}$ *then we claim that*

$$u \in C^\infty(\{x_1^2 + x_2^2 < x_3^2\} \cap \Omega^+).$$

In other words, u *is* C^∞ *smooth inside of the characteristic cone of the future* K_+.

Sketch of the proof. (in the simplest case $f = f(x, u)$). Evidently, the commutator $[\Box, M] = 2\Box$. Inductively we have that $\Box(M^j u) = f_j(x, u, \dots, M^j u)$ for some C^∞ smooth functions $f_j(x, u, \dots, M^j u)$ of their arguments. In fact, $\Box(Mu) = M\Box u + 2\Box u \in H^{s-1}(\Omega)$ according to Lemma 8.1. Then if $M^{j-1} u \in H^s(\Omega) \Rightarrow \Box(M^j u) = M\Box(M^{j-1} u) + 2\Box(M^{j-1} u) \in H^{s-1}(\Omega)$. From $\Box(M^j u) \in H^s(\Omega)$, $M^j u \in H^{s-1}(\Omega^-)$ it follows (see [127]) that $M^j u \in H^s(\Omega)$ (propagation of regularity result to the wave operator $\Box$).

It is an easy exercise on microlocal analysis to prove that if $M^j u \in H^s(\Omega)$ for every j then $u \in C^\infty$ near each point $0 \neq x_0 \in \{x_1^2 + x_2^2 < x_3^2\}$.

In fact, let $(x_0, \xi_0) \in WF(u)$ for some $\xi_0 \neq 0$. Considering the cases $< \xi_0, x_0 > \neq 0$ and $< \xi_0, x_0 > = 0$ we conclude that either M is microlocally elliptic at (x_0, ξ_0) or $\Box$ is microlocally elliptic there. Thus, $u \in C^\infty$ near x_0 and everything is proved.

Proposition 8.2.

Let $u_0 \in H^s(\tilde{\Omega})$, $u_1 \in H^{s-1}(\tilde{\Omega})$, $s > 2.5$ and u satisfies Eq. (8.1), Eq. (8.2). Assume that $\tilde{M}^j u_0 \in H^s(\tilde{\Omega})$ and $\tilde{M}^j u_1 \in H^{s-1}(\tilde{\Omega})$ for each $j \in \mathbf{N}$. Then we claim that $u \in C^\infty(\{x_1^2 + x_2^2 < x_3^2\} \cap \{\tilde{\Omega}^+ \times (0, T)\})$ and $M^j u \in C((0, T); H^s(\tilde{\Omega}^+)) \cap C^1((0, T); H^{s-1}(\tilde{\Omega}^+))$ for each $j \in \mathbf{N}$.

The proof is inductive and standard and we omit it because of the lack of space (see [18]).

Below we propose the following interesting theorem of M. Beals (see [18]) with a detailed proof. On the subject see also [23] and [119].

Theorem 8.1.

Assume that the Cauchy problem Eq. (8.1), Eq. (8.2) has initial data $(u_0, u_1) \in H^s(\tilde{\Omega}) \cap H^{s-1}(\tilde{\Omega})$, $\tilde{\Omega} \subset \mathbf{R}^2$, $s > 2.5$. Let the initial data be radially smooth, i.e. for all $j \in \mathbf{N} : \tilde{M}^j u_0 \in H^s(\tilde{\Omega})$ and $\tilde{M}^j u_1 \in H^{s-1}(\tilde{\Omega})$, $\tilde{M} = x_1 \frac{\partial}{\partial x_1} + x_2 \frac{\partial}{\partial x_2}$. Then we claim that on $\tilde{\Omega}^+ \times (0, T)$ the following inclusion holds:

$$singsupp\, u \subset \{x_1^2 + x_2^2 = x_3^2\} \cup \{(rcos\varphi \mp tsin\varphi, rsin\varphi \pm tcos\varphi, t) :$$

$$r(cos\varphi, sin\varphi) \in singsupp(u_0, u_1)\}.$$

We point out that with the exception of the cone K_+ over 0 the only singularities of the solution u to the Cauchy problem Eq. (8.1), Eq. (8.2) are the same as those for the corresponding linear problem. Put $\beta_{1,2}$:

$$\begin{vmatrix} x_1 = rcos\varphi \mp tsin\varphi \\ x_2 = rsin\varphi \pm tcos\varphi \,. \\ x_3 = t \end{vmatrix}$$

For r, φ fixed these are straight lines through $r(cos\varphi, sin\varphi, 0)$ with mutually orthogonal colinear vectors $(\mp sin\varphi, \pm cos\varphi, 1)$ and generatrices of the cone K (characteristics to $\Box$).

If r describes $\mathbf{R}^1$ and φ is fixed then $\beta_{1,2}$ are characteristic planes passing through the origin: $\beta_{1,2} : x_1 sin\varphi - x_2 cos\varphi \pm x_3 = 0$, $\beta_1 \perp \beta_2$.

Let $x_3 = const \neq 0$. Evidently $\beta_1 \cap \{x_3 = const\}$, $\beta_2 \cap \{x_3 = const\}$ are parallel and are tangential to the circle $x_1^2 + x_2^2 = x_3^2 = const^2$ at the points $\pm(\tilde{x}_1, \tilde{x}_2, x_3 = const)$ (see Fig. 8.1).

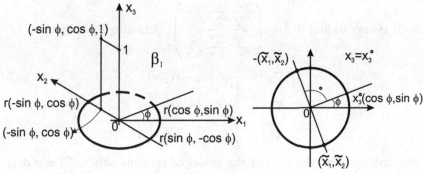

Fig. 8.1

Proof of Theorem 8.1

According to Proposition 8.1 u is C^∞ smooth in the light cone of the future $K_+ : x_3 > \sqrt{x_1^2 + x_2^2}$. Outside of the cone K_+, by finite propagation speed, the values of u are determined by the values of (u_0, u_1) outside from the origin. On the compliment of K_+ in $\{x_3 > 0\}$ we look for a change of the variables after which $\square$ can be written as a sum of M^2 plus second order operator involving two variables only. The corresponding change is

$$x_1 = rs, x_2 = rt, x_3 = rc(s, t) \tag{8.3}$$

and the smooth function c will be found later (see [17] and [18]).

One can easily see that

$$\frac{\partial}{\partial r} = s\frac{\partial}{\partial x_1} + t\frac{\partial}{\partial x_2} + c\frac{\partial}{\partial x_3} = \frac{x_1}{r}\frac{\partial}{\partial x_1} + \frac{x_2}{r}\frac{\partial}{\partial x_2} + \frac{x_3}{r}\frac{\partial}{\partial x_3} = \frac{M}{r} \tag{8.4}$$

$$\frac{\partial}{\partial s} = r\frac{\partial}{\partial x_1} + rc_s\frac{\partial}{\partial x_3}, \frac{\partial}{\partial t} = r\frac{\partial}{\partial x_2} + rc_t\frac{\partial}{\partial x_3}$$

$$M^2 = r^2\partial_r^2 + r\frac{\partial}{\partial r} = r^2\partial_r^2 + M.$$

The Jacobi matrix of the change Eq. (8.4) is

$$J = \begin{pmatrix} \frac{\partial x_1}{\partial r} & \frac{\partial x_1}{\partial s} & \frac{\partial x_1}{\partial t} \\ \frac{\partial x_2}{\partial r} & \frac{\partial x_2}{\partial s} & \frac{\partial x_2}{\partial t} \\ \frac{\partial x_3}{\partial r} & \frac{\partial x_3}{\partial s} & \frac{\partial x_3}{\partial t} \end{pmatrix} = \begin{pmatrix} s & r & 0 \\ t & 0 & r \\ c & rc_s & rc_t \end{pmatrix} \Rightarrow$$

$$D = \det J = r^2(c - sc_s - tc_t) \neq 0.$$

It is easy to find $J^{-1} = \begin{pmatrix} \frac{\partial r}{\partial x_1} & \frac{\partial r}{\partial x_2} & \frac{\partial r}{\partial x_3} \\ \frac{\partial s}{\partial x_1} & \frac{\partial s}{\partial x_2} & \frac{\partial s}{\partial x_2} \\ \frac{\partial t}{\partial x_1} & \frac{\partial t}{\partial x_2} & \frac{\partial t}{\partial x_3} \end{pmatrix}$, obtaining this way that

$$\nabla_x r = \frac{r}{D}(-rc_s, -rc_t, r) \tag{8.5}$$

$$\nabla_x s = \frac{r}{D}(c - tc_t, sc_t, -s)$$

$$\nabla_x t = \frac{r}{D}(tc_s, c - sc_s, -t)$$

We shall work on the level of the principal symbols only as $\Box$ is strictly hyperbolic. Thus, $\Box$ is transformed into a second order linear partial differential operator having the principal symbol

$$c_{11}\frac{\partial^2}{\partial r^2} + c_{22}\frac{\partial^2}{\partial s^2} + c_{33}\frac{\partial^2}{\partial t^2} + 2c_{12}\frac{\partial^2}{\partial r \partial s} + 2c_{13}\frac{\partial^2}{\partial r \partial t} + 2c_{23}\frac{\partial^2}{\partial s \partial t}. \tag{8.6}$$

Having in mind that

$$\frac{\partial^2}{\partial x_i^2} = \left(\frac{\partial r}{\partial x_i}\right)^2 \frac{\partial^2}{\partial r^2} + \left(\frac{\partial s}{\partial x_i}\right)^2 \frac{\partial^2}{\partial s^2} + \left(\frac{\partial t}{\partial x_i}\right)^2 \frac{\partial^2}{\partial t^2} +$$

$$2\frac{\partial r}{\partial x_i}\frac{\partial s}{\partial x_i}\frac{\partial^2}{\partial r \partial s} + 2\frac{\partial r}{\partial x_i}\frac{\partial t}{\partial x_i}\frac{\partial^2}{\partial r \partial t} + 2\frac{\partial s}{\partial x_i}\frac{\partial t}{\partial x_i}\frac{\partial^2}{\partial s \partial t} +$$

first order differential operator in $\frac{\partial}{\partial r}$, $\frac{\partial}{\partial s}$, $\frac{\partial}{\partial t}$ we determine the coefficients c_{ij} of Eq. (8.6).

More specially,

$$c_{12} = 2\frac{\partial r}{\partial x_3}\frac{\partial s}{\partial x_3} - 2\frac{\partial r}{\partial x_1}\frac{\partial s}{\partial x_1} - 2\frac{\partial r}{\partial x_2}\frac{\partial s}{\partial x_2} = \tag{8.7}$$

$$\frac{2r^3}{D^2}(cc_s - s - tc_tc_s + c_t^2 s),$$

$$c_{13} = \frac{2r^3}{D^2}(cc_t - t - sc_tc_s + tc_s^2).$$

Therefore, $c_{12} = c_{13} = 0$ iff the following nonlinear system with respect to $c(s, t)$ holds:

$$cc_s - tc_tc_s + sc_t^2 = s \tag{8.8}$$

$$cc_t - sc_tc_s + tc_s^2 = t.$$

From Eq. (8.8) it follows that

$$tc_s(c - tc_t) + stc_t^2 = sc_t(c - sc_s) + stc_s^2, \tag{8.9}$$

i.e. $(tc_s - sc_t)\frac{D}{r^2} = 0$.

Therefore, $D \neq 0 \iff$

$$tc_s = sc_t. \tag{8.10}$$

The first equation of Eq. (8.8) implies then that $s = cc_s$ and the second one: $t = cc_t$.

Our Ansatz for c is the following one:

$$c^2 = s^2 + t^2 + k, k = const \neq 0.$$

To fix the ideas let $c = \sqrt{s^2 + t^2 + k} \Rightarrow cc_s = s, tc_t = t, \frac{D}{r^2} = c - sc_s - tc_t = \frac{k}{\sqrt{s^2+t^2+k}}$.

As we are working in $x_1^2 + x_2^2 > x_3^2$ the constant k should be negative. Conclusion:

The change Eq. (8.3) satisfies all the necessary conditions and is not degenerate for $c = \sqrt{t^2 + s^2 - 1}$.

Our next steps are to find $c_{11}, c_{22}, c_{33}, c_{23}$. As we know $M^2 = M + r^2 \frac{\partial^2}{\partial r^2}$. After some computations we see that the principal symbol of $\square$ in the new coordinates takes the form (up to a positive smooth factor):

$$-r^2 \frac{\partial^2}{\partial r^2} + (s^2 - 1)\frac{\partial^2}{\partial s^2} + 2st\frac{\partial^2}{\partial s \partial t} + (t^2 - 1)\frac{\partial^2}{\partial t^2}. \tag{8.11}$$

According to Proposition 8.2 we know the regularity of Mu, $M^2 u$. Therefore, we have to study the regularity of the solution of

$$Lu = \left[(s^2 - 1)\frac{\partial^2}{\partial s^2} + 2st\frac{\partial^2}{\partial s \partial t} + (t^2 - 1)\frac{\partial^2}{\partial t^2}\right]u = g(s, t, r, u, \nabla u, u_{rr}), r > 0, \tag{8.12}$$

$$g \in C^\infty$$

with respect to its arguments.

In fact, for $r > 0 : r^2 u_{rr} = (M^2 - M)u$. As it is shown in [17], the singularities of Eq. (8.12) propagate as in the 1D case, $r > 0$ being a parameter. On the other hand it is shown in [118] that there are no new born nonlinear singularities for a second order equation in 1D. Consequently, the singularities we are interested in are propagating along the characteristics of the linear operator L in the left-hand side of Eq. (8.12). The polar change

$$s = acosb \tag{8.13}$$

$$t = asinb$$

in L and the change formulas

$$\frac{\partial}{\partial s} = cosb\frac{\partial}{\partial a} - \frac{sinb}{a}\frac{\partial}{\partial b} \qquad (8.14)$$

$$\frac{\partial}{\partial t} = sinb\frac{\partial}{\partial a} + \frac{cosb}{a}\frac{\partial}{\partial b}, s\frac{\partial}{\partial s} + t\frac{\partial}{\partial t} = a\frac{\partial}{\partial a}$$

reduces L modulo lower order terms to the operator $\tilde{L}$ in $\mathbf{R}^2$ with principal symbol $(a^2 - 1)\frac{\partial^2}{\partial a^2} - \frac{1}{a^2}\frac{\partial^2}{\partial b^2}$, $a > 0$. Without loss of generality we shall deal with the 1D hyperbolic operator $(a^2 - 1)a^2\frac{\partial^2}{\partial a^2} - \frac{\partial^2}{\partial b^2} = \tilde{L}$. Our last change in $\tilde{L}$ is $\tilde{a} = arcseca$, $b = b$. We remind of the reader that $seca = \frac{1}{cosa}$, $(seca)' = +\frac{sina}{cos^2a} = secatga$,

$$(arcsecx)' = \frac{1}{x\sqrt{x^2 - 1}}, 0 < arcsecx < \frac{\pi}{2},$$

$$(arcsecx)' = -\frac{1}{x\sqrt{x^2 - 1}}, \frac{\pi}{2} < arcsecx < \pi,$$

$$\left(arcsec\sqrt{x^2 + 1}\right)' = (arctgx)'\varepsilon(x), \varepsilon^2(x) = 1$$

$$arctgx \pm arctgy = arctg\frac{x \pm y}{1 \mp xy}.$$

Making the change $\tilde{a} = arcseca$ in $\tilde{L}$ we obtain the operator $\tilde{\tilde{L}}$ with principal part $\frac{\partial^2}{\partial \tilde{a}^2} - \frac{\partial^2}{\partial b^2}$, i.e. the string operator. Its characteristics are

$$\tilde{a} = \pm b + \alpha, \qquad (8.15)$$

α being the parameter of the corresponding straight lines. We are going back now from the coordinates $(\tilde{a}, b)$ to the "old" coordinates (x_1, x_2, x_3) in order to write the characteristics of $\tilde{\tilde{L}}$ in the initial coordinates. Thus, Eq. (8.15) takes the form

$$sec\left(\pm arctg\frac{x_2}{x_1} + \alpha\right) = \frac{\sqrt{x_1^2 + x_2^2}}{\sqrt{x_1^2 + x_2^2 - x_3^2}}, \qquad (8.16)$$

as $a = \sqrt{s^2 + t^2}$, $a = sec(\pm b + \alpha)$, $b = arctg\frac{x_2}{x_1}$, $\sqrt{x_1^2 + x_2^2} = r\sqrt{s^2 + t^2}$, $x_3^2 = r^2(s^2 + t^2) - r^2 \Rightarrow r^2 = x_1^2 + x_2^2 - x_3^2$. If $x_3 = 0$ we get that $\alpha = \mp arctg\frac{x_2}{x_1}$. Assume moreover, that the point $(rcos\varphi, rsin\varphi, 0) \in singsuppu_0 \cup singsuppu_1$ belongs to the family of characteristics Eq. (8.16). Then $\alpha = \mp\varphi$.

As we mentioned before $r = \sqrt{x_1^2 + x_2^2 - x_3^2}$ is considered as positive parameter. Put $p = x_3 = \sqrt{x_1^2 + x_2^2 - r^2}$. Then Eq. (8.16) implies that

$$arcsec\sqrt{\frac{p^2 + r^2}{r^2}} = \pm\left(arctg\frac{x_2}{x_1} - \varphi\right). \qquad (8.17)$$

The trigonometric identities formulated above give us that $arctg\frac{p}{r} = \pm(arctg\frac{x_2}{x_1} - \varphi) = \pm arctg(\frac{\frac{x_2}{x_1} - tg\varphi}{1 + \frac{x_2}{x_1}tg\varphi})$, i.e.

$$\frac{x_2}{x_1} = \frac{\pm\frac{p}{r} + tg\varphi}{1 \mp \frac{p}{r}tg\varphi} = \frac{\pm p cos\varphi + r sin\varphi}{r cos\varphi \mp p sin\varphi}. \qquad (8.18)$$

This way we conclude that the two characteristics (straight lines) through the singular point $(rcos\varphi, rsin\varphi, 0)$ are given by $(rcos\varphi \mp psin\varphi, \pm pcos\varphi + rsin\varphi, p)$. Theorem 8.1 is proved with parameter $p \in \mathbf{R}$ instead of t as it is in the inclusion for *singsuppu*.

Examples. We can deal with the functions x^α, $\alpha > 1$ defined as $|x|^\alpha$ or $\bar{x}^\alpha = \begin{cases} |x|^\alpha, & x \geq 0 \\ -|x|^\alpha, & x < 0 \end{cases}$. Certainly, $\alpha \notin \mathbf{N}$ and α could be arbitrary large. Let

$$u_i = \sum_{k=1}^N a_{ik}|x_1|^{\alpha_{ik}}\bar{x}_2^{\beta_{ik}}, \alpha_{ik} > 1, \beta_{ik} > 1, \alpha_{ik}, \beta_{ik} \notin \mathbf{N},$$

$\alpha_{ik} + \beta_{ik} = m_i > 1$ for $i = 0, 1$. The initial data are finitely smooth and homogeneous of order m_i.

Therefore, $\tilde{M}u_i = m_iu_i$. Evidently, $u_i(\rho, \Theta) = \rho^{m_i}\sum_{k=1}^N a_{ik}|cos\Theta|^{\alpha_{ik}}$ $sgnx_2|sin\Theta|^{\beta_{ik}} = v_i(\rho)\Phi_i(\Theta)$. In general, $\{x_1 = 0\} \times \{x_2 = 0\} \subset$ $singsuppu_i$, i.e. $\Theta = 0, \pi$ and $\Theta = \frac{\pi}{2}, \frac{3}{2}\pi$ there.

In more general situation $u_i = v_i(\rho)\Phi_i(\Theta)$, $i = 0, 1$, v_i is $C^{[k]}$ smooth, $k \gg 1$, but not $C^{[k]+1}$ and the same with respect to $\Phi_i(\Theta)$. Usually, $v_i \in C^\infty(\rho > 0)$.

Denote by C a Cantor set in $[0, 2\pi]$ and suppose that $\Phi_i(\Theta)$ are singular (i.e. not C^∞ smooth) on C. As we know C has the cardinality of the continuum, is nowhere dense in $[0, 2\pi]$ and its Lebesgue measure is 0. Then the singularity of the solution u of Eq. (8.1), Eq. (8.2) will propagate as in the linear case. Below we propose Fig. 8.2 as a geometrical visualisation. Theorem 8.1 can be proved by applying the results of [119] too.

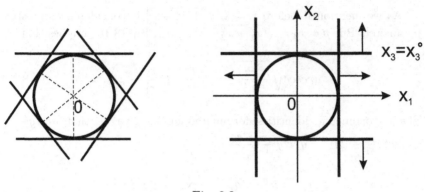

Fig. 8.2

Straight lines (two by two) are parallel and tangential to the circle $x_1^2 + x_2^2 = (x_3^0)^2$.

Remark 8.1. We propose here an interesting result of Bony [23], [25]. Recall that a distribution $u \in H_{loc}^s$ is called conormal with respect to the nondegenerate surface Σ if $M_1 M_2 \ldots M_j u \in H_{loc}^s$ for all smooth vector fields $M_1, \ldots, M_j$ tangential to Σ.

Theorem 8.2. *(Bony)*
Let $u \in H_{loc}^s(\Omega)$, $\Omega \subset \mathbf{R}^3$, $s > \frac{5}{2}$ and u satisfies Eq. (8.1). Assume that u is conormal with respect to $\Sigma = (\Sigma_1, \Sigma_2, \Sigma_3)$ in Ω^-, where Σ_1, Σ_2, Σ_3 are characteristic hyperplanes of $\Box$ intersecting transversally at the origin. Then

$$u \in C^\infty \left(\Omega \setminus \Sigma_1 \cup \Sigma_2 \cup \Sigma_3 \cup \left\{ x_3 = \sqrt{x_1^2 + x_2^2} \right\} \right).$$

Geometrically we have Fig. 8.3

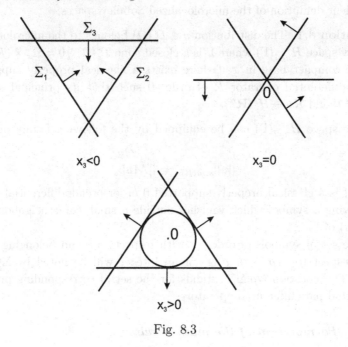

Fig. 8.3

Thus, 3 progressing waves are propagating for $x_3 < 0$. At the moment $x_3 = 0$ they collide and due to their interaction new singularity was born for $x_3 > 0$. It propagates along the light cone of future $K_+ : x_3 = \sqrt{x_1^2 + x_2^2}$. Certainly, the progressing waves are continuing their propagation for $x_3 > 0$. The light cone is inscribed in the pyramid with vertex at 0 that is generated by Σ_1, Σ_2, Σ_3 for $x_3 > 0$. It is interesting to know that the new born singularity is weaker in the Sobolev scale of spaces than the singularities carried out by the three progressing waves [25].

8.3 Wave fronts of the solutions of fully nonlinear symmetric positive systems of PDE

1. Suppose that the definition and the main properties of the paradifferential operators are well known to the reader (see for example [24]). The symbol $C^\rho(\Omega)$ stands for the classical Hölder spaces in the domain $\Omega \subset \mathbf{R}^n$ and $\rho > 0$ in not an integer. We will assume that only real-valued solutions $u \in H^s_{loc}(\Omega)$ will be studied. We shall denote by $||.||_s$ Sobolev's norm

in H^s and by $(,)$ the L_2 complex scalar product. Below we give another equivalent definition of the microlocalized Sobolev spaces.

Definition 8.1. The distribution $u \in D'(\Omega)$ belongs to the microlocalized Sobolev space $H^s_{mcl}(\Gamma)$ where Γ is a closed cone $T^*_x(\Omega) \setminus 0 = \Omega_x \times (\mathcal{R}^n_\xi \setminus 0)$ with a compact base in x if there exists a classical properly supported pseudodifferential operator $\mathcal{R}$ of order 0 such that its principal symbol $\mathcal{R}^0|_\Gamma \neq 0$ and $\mathcal{R}u \in H^s(\mathbf{R}^n)$.

The space $H^s_{mcl}(\Gamma)$ can be equipped by the following family of seminorms:

$$\|u\|_{s,mcl(\Gamma)} = \|\mathcal{M}u\|_s$$

and $\mathcal{M}$ is a classical properly supported 0 order pseudodifferential operator having a symbol which vanishes outside a small conic neighbourhood (ngbhd) of Γ.

The set of symbols of order m with respect to ξ and belonging to C^ρ with respect to $x, \rho > 0$, ρ is not an integer will be noted by Σ^m_ρ (see [24]). The notation $Op(\Sigma^m_\rho)$ stands for the set of corresponding properly supported paradifferential operators.

8.3.1   *Formulation of the main results*

Assume that $F_k(x, u, p)$, $u = (u_1, \ldots, u_N)$, $p = (u_{11}, \ldots, u_{ij}, \ldots, u_{N_n})$, $1 \leq j \leq n, 1 \leq i, k \leq N$ are real-valued C^∞ functions of their arguments $x \in \Omega$, $u \in \mathbf{R}^n$, $p \in \mathbf{R}^{nN}$ and let $u \in H^s_{loc}(\Omega), s > 3 + n/2$ be a real valued (i.e. $u \in C^{3+\varepsilon}(\Omega), 0 < \varepsilon$ according to Sobolev's embedding theorem) solution of the nonlinear system of PDE

$$F_k(x, u(x), \partial u(x)) = 0, 1 \leq k \leq N, x \in \Omega \qquad (8.19)$$

$$u(x) = (u_1(x), \ldots, u_N(x)), \partial = (\partial_{x_1}, \ldots, \partial_{x_n}).$$

Then the first variation of Eq. (8.19) can be written as

$$\mathcal{P}v = \sum_{j=1}^n A_j(x) D_j v - iB(x)v, \xi_j \leftrightarrow D_j = \frac{1}{i}\partial_{x_j}, \qquad (8.20)$$

where the matrix

$$A_j(x) = \|\frac{\partial F_k}{\partial u_{ij}}(x, u(x), \partial u(x))\|_{1 \leq i,k \leq N} \in C^{2+\varepsilon}$$

and

$$B(x) = \|\frac{\partial F_k}{\partial u_i}(x, u(x), \partial u(x))\|_{1 \leq i,k \leq N} \in C^{2+\varepsilon}, \varepsilon > 0.$$

Define now the matrix

$$K(x) = \frac{B + B^*}{2} - \frac{1}{2}\sum_{j=1}^{n}\frac{\partial A_j(x)}{\partial x_j}.$$

B^* being the Hermitian adjoint matrix of B. As usual the characteristic set of Eq. (8.20) is given by

$$Charp_1 = \left\{\rho = (x, \xi) \in T^*(\Omega) \setminus 0 : \det \sum_{j=1}^{n} A_j(x)\xi_j = 0\right\}.$$

Theorem 8.3. *(see [105])*
  Consider the fully nonlinear system Eq. (8.19) and suppose that the real-valued solution $u \in H^s_{loc}(\Omega), s > 3 + n/2$ is such that

$$(i) \qquad A_j(x) = A_j^*(x), \forall x \in \Omega,$$

(i.e. $\frac{\partial F_k}{\partial u_{ij}}(x, u(x), \partial u(x)) = \frac{\partial F_i}{\partial u_{kj}}(x, u(x), \partial u(x)) \; \forall x \in \Omega$).
  We will assume that $\rho^0 = (x^0, \xi^0) \in Charp_1, |\xi^0| = 1, \Gamma$ is a conical neighbourhood of ρ^0 and such that $u \in H^t_{mcl}(\Gamma \setminus \rho^0),$[1] $t < 2s - 2 - n/2$. Then if

$$(ii)_0 \qquad K(x^0) > \gamma Id_N, \gamma = const > 0,$$

$$(iii)_t \qquad K(x^0) + t\sum_{k,j=1}^{n}\frac{\partial A_j}{\partial x_k}(x^0)\xi_k^0\xi_j^0 > \gamma Id_N,$$

are valid we claim that $u \in H^t_{mcl}(\rho^0)$.

  Certainly, Id_N is the identity matrix of $\mathbf{R}^N$.

Remark 8.2. Obviously $K(x)$ and the matrix participating in the left-hand side of $(iii)_t$ are symmetric due to (i). It is easy to verify that if $0 < \lambda < t$ then $(iii)_t$ implies $(iii)_\lambda$. The restriction on the strength of the singularity $t < 2s - 2 - n/2$ is natural in the framework of Bony's theory (see [24]) and can be explained by the fact that this theory is valid "under the shock" only.

---

[1] I.e. $\mathcal{M}u \in H^t(\mathbf{R}^n)$ for each properly supported classical 0 order pseudodifferential operator and such that its conical support cone $supp\mathcal{M} \subset\subset \Gamma \setminus \rho^0$.

Corollary 8.1.

Let the solution u of (19) belong to $H^s(x_0)$, $s > 3 + n/2$, i.e. $u \in H^s$ in a small neighbourhood of x_0. Moreover, suppose that the relation $u \in H^s_{mcl}(\rho)$, $\rho = (x^0, \xi)$, $|\xi| = 1$ be eventually violated in finitely many characteristic points and $u \in H^t(x)$ for all $x \neq x^0$ in a small neighbourhood of x^0. Then (i) and the fulfillment of $(ii)_0$ and $(iii)_t$ at these points imply that $u \in H^t(x^0)$.

Remark 8.3. It is interesting to note that $u \in H^{2s-1-\varepsilon-n/2}(\rho^0)$, $\varepsilon > 0$ for each point $\rho^0 \notin Charp_1$ (see Th. 5.4 from [24]).

Standard considerations from the theory of the paradifferential operators $\mathcal{P} \in Op(\Sigma^1_\sigma)$, $\sigma > 2$, σ is not an integer [24] reduce the proof of Theorem 8.3 to the proof of the next assertion (see [105]).

Theorem 8.4.

Consider the first order paradifferential system

$$\mathcal{P}(x,D)u = \sum_{j=1}^n A_j(x)D_j u - iB(x)u = f = (\mathcal{P}_1 - iB)u, \qquad (8.21)$$

$$(-\mathcal{P}(x,D)u = -f),$$

where $\mathcal{P} \in Op(\Sigma^1_\sigma)$, $\sigma > 2$ is not an integer, $A_j^*(x) = A_j(x) \forall x \in \Omega$; $A_j(x)$, $B(x)$ are real-valued $N \times N$-matrices and the conditions $(ii)_0$ and $(iii)_t$, $t > 2$ are fulfilled.

Then $u \in H^{t-1/2}_{comp}(\Omega)$, $u \in H^t_{mcl}(\Gamma \setminus \rho^0)$, $f \in H^t_{mcl}(\Gamma)$, where $\rho^0 \in Charp_1$ and Γ is a conic neighbourhood of $\rho^0 = (x^0, \xi^0)$ imply that $u \in H^t_{mcl}(\rho^0)$, i.e. $u \in H^t_{mcl}(\Gamma)$.

Assume now that Theorem 8.4 is proved and $s < t < 2s - 2 - n/2$ (the non-trivial case). In order to verify Theorem 8.3 we apply Theorem 5.3.b) from [24] with the corresponding notations $d = 1$, $\rho = s - \varepsilon - n/2$, $1 \gg \varepsilon > 0$, $\sigma = \rho - 1$ and we conclude that one can find such paradifferential operator $\mathcal{P} \in Op(\Sigma^1_\sigma)$, $\sigma > 2$ given by Eq. (8.21) that

$$\mathcal{P}u \in H^{2s-2-\varepsilon-n/2}_{loc} \Rightarrow \mathcal{P}u \in H^t_{loc} (\text{as } 0 < \varepsilon \ll 1).$$

There are no difficulties to verify that if $u \in H^s_{loc}$, $\mathcal{P}u \in H^t_{mcl}(\Gamma)$, $u \in H^{t-1/2}_{mcl}(\Gamma)$, $u \in H^t_{mcl}(\Gamma \setminus \rho^0)$ then $u \in H^t_{mcl}(\Gamma)$. To do this consider the classical pseudodifferential operator $\mathcal{J} \in S^0_{1,0}$, $\mathcal{J} \equiv 1$ in a small conic ngbhd of Γ, $\mathcal{J} \equiv 0$ outside a larger conic ngbhd of Γ. Then $\mathcal{J}u \in H^{t-1/2}_{comp}$, $\mathcal{J}u \in$

$H_{mcl}^t(\Gamma \setminus \rho^0)$ and $\mathcal{P}(\mathcal{J}u) \in H_{mcl}^t(\Gamma)$, as $\mathcal{P}(I - \mathcal{J})u \in H_{mcl}^t(\Gamma)$ according to Corollary 3.5 from [24]. So $\mathcal{J}u \in H_{mcl}^t(\Gamma)$ and therefore $u \in H_{mcl}^t(\Gamma)$. To complete the proof of Theorem 8.3 on the base of Theorem 8.4 we use standard bootstrap arguments (see for example the proof of Proposition 26.6.1 from [67]) as well as the fact that if $0 < \lambda < t$ then $(iii)_\lambda$ is fulfilled. We omit the details.

Remark 8.4. Consider the same system Eq. (8.19) on the n-dimensional torus T^n. i.e. $x \in T^n$ and the solutions of Eq. (8.19) belong to $H^s(T^n)$. It was shown by Moser (see [98]) that a classical $C^2(T^n)$ solution exists. Suppose now that

$$\frac{\partial F_k}{\partial u_{ij}}(x, u, p) = \frac{\partial F_i}{\partial u_{kj}}(x, u, p), \tag{8.22}$$

$\forall x \in T^n$ and for each (u, p) such that $|u| + |p| < 1$. Let $u \equiv 0$ be a solution of Eq. (8.19), Eq. (8.22) is valid and $(ii)_0$, $(iii)_t$, $t > 3 + n/2$, are fulfilled for each $x \in T^n$, $\forall \xi$, $|\xi| = 1$, $u = p = 0$. Then there exists $s_1 > 3 + n/2$ with the property: for every $s \in (3 + n/2, s_1)$ one can find such a ngbhd $W \ni 0$ in $H^s(T^n)$ that the system $F_k(x, u, \partial u) = f_k$ is solvable in $H^s(T^n)$ for arbitrary $f \in W$. For the proof see [12].

2. We shall investigate here the propagation of regularity to the solutions of symmetrizable hyperbolic systems and shall illustrate it by an example from fluid mechanics. Consider the Cauchy problem

$$\frac{\partial u}{\partial t} + \sum_{j=1}^n A_j(u)\frac{\partial u}{\partial x_j} + B(u) = 0, u|_{t=0} = u_0(x) \tag{8.23}$$

in the strip $(t, x) \in [0, T] \times \mathbf{R}_x^n$. Here $u = \begin{pmatrix} u_1 \\ \cdot \\ \cdot \\ \cdot \\ u_N \end{pmatrix}$, $A_j(u)$ and $B(u)$ are real-valued matrices of the type $N \times N$, respectively $N \times 1$, C^∞ smooth in some open domain $\Omega \subset \mathbf{R}^N$. The system Eq. (8.23) is supposed to be hyperbolic symmetrisable, i.e. the following conditions hold

(H) (iv) There exists real symmetric C^∞ smooth in Ω matrix S and such that

(iv_a) $S(u) \geq S_0 I_N, S_0 > 0, \forall u$

(iv_b) $S(u)A_j(u), j = 1, \ldots, n$ are symmetric.

The system is hyperbolic as the eigenvalues of $\sum_{j=1}^n A_j(u)\xi_j$ are real for $\xi \in \mathbf{R}^n$ and $u \in \Omega$.

Due to (H) the solvability of Eq. (8.23) and the regularity of its solution are reduced to the investigation of the symmetric hyperbolic system

$$S(u)\frac{\partial u}{\partial t} + \sum_{j=1}^{n} S(u)A_j(u)\frac{\partial u}{\partial x_j} + SB(u) = 0, u|_{t=0} = u_0(x). \qquad (8.24)$$

The matrix $S(u)B(u)$ is not obliged to be symmetric.

Evidently, Eq. (8.24) satisfies the condition (i). (i) seems to be artificial as it depends on the unknown solution u and we are going to study the regularity of u via u.

Theorem 8.5. [12]

Assume that the system Eq. (8.23) verifies the condition (H), $u_0 \in H^s$, $s > \frac{n}{2} + 1$. Then there exists $T > 0$ and solution $u \in L^\infty([0,T]; H^s) \cap Lip([0,T]; H^{s-1})$ of Eq. (8.23).

Moreover, $u \in C([0,T]; H^s) \cap C^1([0,T]; H^{s-1})$.

Therefore, we have both local in time existence and regularity result to Eq. (8.23). The natural question is whether systems of the form Eq. (8.23)-Eq. (8.24) appear in the applications of PDE.

The positive answer of that question is given by the system

$$\frac{\partial \rho}{\partial t} + div(\rho v) = 0 \qquad (8.25)$$

$$\frac{\partial(\rho v_i)}{\partial t} + div(\rho v_i v + p e_i) = 0, i = 1,\ldots,n,$$

describing the state of compressible fluid in $\mathbf{R}^n$ with density $\rho > 0$, velocity $v = (v_1, \ldots, v_n)$ and given pressure $p(\rho) \in C^\infty$, $p'(\rho) > 0$. The vector e_i from Eq. (8.25) is equal to $(0, \ldots, 0, 1, 0, \ldots 0)$ with 1 at the i-th position and the entropy of the system is constant. Evidently, $B \equiv 0$, $N = n + 1$.

After some calculations Eq. (8.25) can be written in the equivalent form

$$\frac{\partial \rho}{\partial t} + div(\rho v) = 0 \qquad (8.26)$$

$$\frac{\partial v_i}{\partial t} + <v, \nabla v_i> + \frac{c^2}{\rho}\frac{\partial \rho}{\partial x_i} = 0, i = 1,\ldots,n,$$

where $c^2 = p'(\rho) > 0$; $c > 0$, $c \in C^\infty(\mathbf{R}_+^1)$; $(t, x) \in \mathbf{R}^{n+1}$.

Put $u = (\rho, v) \in \mathbf{R}^{n+1}$. Then Eq. (8.26) is equivalent to

$$\frac{\partial u}{\partial t} + \sum_{j=1}^{n} A_j(u)\frac{\partial u}{\partial x_j} = 0. \qquad (8.27)$$

One can easily see that $A_j(u)$ is a $(n+1) \times (n+1)$ nonsymmetric matrix, given by the formula

$$A_j(u) = \begin{pmatrix} v_j & \cdots & \rho & \cdots \\ \vdots & & \vdots & \\ \cdots & v_j & & \cdots \\ c^2/\rho & \cdots & v_j & \cdots \\ \vdots & & \cdots\cdots & v_j \end{pmatrix}.$$

The other elements of A_j are 0. On the diagonal stands v_j, c^2/ρ stands on the $(j+1)$ row and ρ stands on the $(j+1)$ column.

Consider now the diagonal $(n+1) \times (n+1)$ matrix

$$S(u) = \begin{pmatrix} c^2 & & & \\ & \rho^2 & & \\ & & \ddots & \\ & & & \rho^2 \end{pmatrix}.$$

Standard computations show that SA_j is symmetric matrix for $j = 1, \ldots, n$. In fact,

$$SA_j = \begin{pmatrix} c^2 v_j & \cdots & c^2\rho & \cdots \\ \vdots & \rho^2 v_j & \vdots & \\ c^2\rho & \cdots & \rho^2 v_j & \cdots \\ \vdots & & \cdots & \rho^2 v_j \end{pmatrix}.$$

Assume now that $u \in H^s_{loc}$, $s > \frac{n+1}{2} + 1$ satisfies Eq. (8.26), i.e. Eq. (8.27). According to Bony's theorem 5.3.b) from [24], the corresponding to Eq. (8.27) paradifferential operator $\tilde{\mathcal{P}} \in Op(\Sigma^1_\lambda)$, $\lambda = s - \frac{n+1}{2} - \varepsilon$, $0 < \varepsilon \ll 1$, $\lambda > 1$, $\lambda \notin \mathbf{N}$, $u \in C^\lambda$. The principal symbol of $\tilde{\mathcal{P}}$ is $\tilde{p}_1 = S(u)\tau + \sum_{j=1}^n S(u)A_j\xi_j$, $det p_1 = det S.det(\tau + \sum_{j=1}^n A_j\xi_j)$, i.e. $Char\tilde{p}_1$ is the same as $Char p_1$.

Proposition 8.3. *(see* [106]*)*

   Consider the symmetric hyperbolic system Eq. (8.24). Then under the assumptions $u \in H^s_{loc}$, $s > \frac{n+3}{2}$ and $u \in H^{t_1}_{mcl}(\rho^0)$, $\forall \rho^0 \in Char p_1 \cap \partial W \cap \{t \geq \delta\}$, where $t_1 < 2s - \frac{n+3}{2}$ and W is an open cone in $T^*(X)$, $X \subset \mathbf{R}^{n+1}_{t,x}$, it follows that

$$u \in H^{t_1}_{mcl}(\rho^0), \forall \rho^0 \in Char p_1 \cap W \cap \{t \geq \delta\}.$$

As we know, $\rho^0 \notin Charp_1 \Rightarrow u \in H_{mcl}^{2s-\frac{n+1}{2}-\varepsilon}$, $0 < \varepsilon \ll 1$ (in the case of microlocal ellipticity we have gain of regularity equal to 1). The above mentioned result can be formulated in a similar way: If there are no H^{t_1} singularities of the solution of Eq. (8.24) on the northern hemisphere there are no H^{t_1} singularities of u on the northern semiball too (see Fig. 8.4).

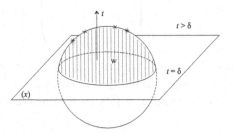

Fig. 8.4. $u \in H_{loc}^s$, $s > \frac{n+3}{2}$, $t_1 < 2s - \frac{n+3}{2}$, $2s - \frac{n+3}{2} > s$, i.e. the possible microlocal gain of regularity is $s - \frac{n+3}{2}$.

$\times$ are the characteristic points on the northern hemisphere and $*$ are the elliptic points there. If u is $H_{mcl}^{t_1}$ smooth there it turns out to be $H_{mcl}^{t_1}$ smooth in the northern semiball too. We shall omit the proof of Proposition 8.3 but will prove Theorem 8.4 as it is more complicated.

8.3.2   *Proof of Theorem 8.4*

When proving Theorem 8.4 we shall have to compute the Poisson bracket $\{p_1, c_1\}$, where $c_1 = k_1(x)\gamma_1(\xi)(1 + \mu^2|\xi|^2)^{-\delta}$, $\delta > 0$, $\mu \in (0, 1]$, $k_1 \in C_0^\infty$, $k_1 \equiv 1$ in a ngbhd of x_0, $\gamma_1 = |\xi|^r h_1(\xi)$, $ord_\xi h_1 = 0$, $h_1 \in C^\infty(\mathbf{R}^n \setminus 0)$, $h_1 \equiv 1$ in a conic ngbhd of ξ^0. As p_1 is the matrix valued symbol of $\mathcal{P}_1$ we can regard c_1 as a diagonal matrix in $\mathbf{R}^n$.

After some calculations we get

$$\{p_1, c_1\} = \sum_{j=1}^n A_j \frac{\partial k_1}{\partial x_j}(1 + \mu^2|\xi|^2)^{-\delta}\gamma_1(\xi) - \sum_{j,k=1}^n r\frac{\partial A_j}{\partial x_k}\xi_j\xi_k h_1|\xi|^{r-2} \times$$

$$k_1(1 + \mu^2|\xi|^2)^{-\delta} - \sum_{j,k=1}^n \frac{\partial A_j}{\partial x_k}\xi_j|\xi|^r\frac{\partial h_1}{\partial \xi_k}k_1(1 + \mu^2|\xi|^2)^{-\delta} +$$

$$2\delta \sum_{j,k=1}^n \frac{\partial A_j}{\partial x_k}\mu^2\xi_j\xi_k k_1\gamma_1(1 + \mu^2|\xi|^2)^{-\delta-1} \equiv I_1 + I_2 + I_3 + I_4.$$

Having in mind that $(1 + \mu^2|\xi|^2)^{-\delta}$ is a bounded family in Σ_λ^0 with respect to μ for each $\lambda > 0$, λ-non-integer and that $\partial k_1 / \partial x_j \equiv 0$ in a ngbhd of x^0, $\partial h_1 / \partial \xi_k \equiv 0$ in a conic ngbhd of ξ^0 we obtain: I_1, I_3 are bounded in $\Sigma_{\sigma-1}^r$, cone $supp(I_1, I_3) \subset\subset \Gamma \setminus \rho^0$. It is trivial to see that

$$\{p_1, c_1\} = I_1 + I_3 + (1 + \mu^2|\xi|^2)^{-\delta} \sum_{j,k=1}^n \frac{\partial A_j}{\partial x_k} k_1 \gamma_1 \xi_j \xi_k \times$$

$$\left(-r|\xi|^{-2} + \frac{2\delta\mu^2}{1 + \mu^2|\xi|^2} \right).$$

Introduce now a cut-off symbol $\eta(x, \xi)$, $ord_\xi \eta = 0$, cone $supp\eta \subset\subset \Gamma$, $\eta \equiv 1$ in a tiny conic ngbhd of ρ^0. Then

$$\{p_1, c_1\} = \eta^2\{p_1, c_1\} + (1 - \eta^2)\{p_1, c_1\}.$$

The observations that $\mu^2 \xi_j (1 + \mu^2|\xi|^2)^{-1}$, $\mu \in (0, 1]$ is a bounded family of symbols in Σ_λ^{-1}, $\forall \lambda > 0$ and $\eta^2 I_1$, $\eta^2 I_3 \equiv 0$ imply that

$$\{p_1, c_1\} = \eta^2 c_1 |\xi|^{-2} \sum_{j,k=1}^n \frac{\partial A_j}{\partial x_k} \xi_j \xi_k \left(-r + \frac{2\delta\mu^2|\xi|^2}{1 + \mu^2|\xi|^2} \right) + (1 - \eta^2) A_r, \quad (8.28)$$

where $A_r = \{p_1, c_1\}$, cone $suppA_r \subset\subset \Gamma$ and A_r is a bounded family of symbols in $\Sigma_{\sigma-1}^r$.

Consider now the symbol

$$c_\mu^2 = k^2(x) h^2(\xi) |\xi|^{2t} (1 + \mu^2|\xi|^2)^{-2} \quad \text{(i.e. } c_\mu \in \Sigma_{\sigma-1}^{t-2}) \quad (8.29)$$

where k and h have the same properties as k_1 and h_1, $\delta = 2$, $r = 2t$.

Then the following decomposition is valid (see Eq. (8.28))

$$\{p_1, c_\mu^2\} = \eta^2 c_\mu^2 |\xi|^{-2} \sum_{j,k=1}^n \frac{\partial A_j}{\partial x_k} \xi_j \xi_k \left(-2t + \frac{4\mu^2|\xi|^2}{1 + \mu^2|\xi|^2} \right) + \quad (8.30)$$

$$(1 - \eta^2) A_{2t} = \eta^2 c_\mu^2 I + (1 - \eta^2) A_{2t}.$$

In Eq. (8.30) A_{2t} is bounded in $\Sigma_{\sigma-1}^{2t}$, cone $suppA_{2t} \subset\subset \Gamma$, I is bounded in $\Sigma_{\sigma-1}^0$ and moreover

$$t > t - \frac{2\mu^2|\xi|^2}{1 + \mu^2|\xi|^2} > t - 2 > 0.$$

We go back to the proof of Theorem 8.4 and consider the bilinear from $Im(c_\mu u, c_\mu f)$, $f = \mathcal{P}u$, which is well defined as $f \in H_{mcl}^t(\Gamma)$, cone

$suppc_\mu \subset\subset \Gamma \Rightarrow c_\mu f \in H^2_{comp}$, $c_\mu u \in H^{3/2}_{comp}$. Having in mind that $p = p_1 - iB$ and c_μ is bounded in $\Sigma^t_{\sigma-1}$ we have

$$Im(c_\mu u, c_\mu f) = Im(c_\mu u, c_\mu p_1 u) + Im(c_\mu u, -ic_\mu Bu).$$

From now on we shall denote by $d_k > 0$ constants which do not depend on $\mu \in (0, 1]$.

So

$$|(c_\mu u, c_\mu f)| \le ||c_\mu u||_0 ||c_\mu f||_0 \le \frac{\varepsilon}{2}||c_\mu u||_0^2 + C(\varepsilon)||f||_{t,mcl(\Gamma)}^2, \forall \varepsilon > 0.$$

The identity $c_\mu u = \eta c_\mu u + (1 - \eta)c_\mu u$ and the fact that c_μ are bounded in $\Sigma^t_{\sigma-1}$, $u \in H^t_{mcl}(\Gamma \setminus \rho^0)$, cone $supp((1 - \eta^2)c_\mu) \subset\subset \Gamma' \subset\subset \Gamma$, $\rho^0 \notin \Gamma'$ and Γ' is a closed cone imply

$$|(c_\mu u, c_\mu f)| \le \varepsilon||\eta c_\mu u||_0^2 + C(\varepsilon)||f||_{t,mcl(\Gamma)}^2 + d_1||u||_{t,mcl(\Gamma')}^2, \forall \varepsilon > 0. \quad (8.31)$$

Evidently,

$$Im(c_\mu u, c_\mu p_1 u) = Im(c_\mu u, p_1 c_\mu u) + Im(c_\mu u, [c_\mu, p_1]u). \quad (8.32)$$

Put $v = c_\mu u \in H^{3/2}_{comp}$ and apply Theorem 3.3 from [24] to the expression $(v, \mathcal{P}_1 v) = (\mathcal{P}_1^* v, v)$. As the symbol of the L_2 adjoint operator $\mathcal{P}_1^*$ of $\mathcal{P}_1$ is given by the formula

$$p_1^* = p_1 - i \sum_{|\alpha|=1} p_{1(\alpha)}^{(\alpha)} + \mathcal{R}, \mathcal{R} \in \Sigma^{-1}_{\sigma-2}, p_{1(\alpha)}^{(\alpha)} = A_\alpha(x)$$

we conclude that

$$Im(v, \mathcal{P}_1 v) = \frac{1}{2i}((\mathcal{P}_1^* - \mathcal{P}_1)v, v) = -\frac{1}{2}\sum_{j=1}^n \left(\frac{\partial A_j}{\partial x_j}v, v\right) + O(||v||_{-1/2}^2),$$

i.e.

$$Im(c_\mu u, p_1 c_\mu u) = -\frac{1}{2}\sum_{j=1}^n \left(\frac{\partial A_j}{\partial x_j}c_\mu u, c_\mu u\right) + O(||u||_{t-1/2}^2), \quad (8.33)$$

and the remainder O is independent of μ.

In a similar way we get

$$Im(c_\mu u, -ic_\mu Bu) = Re(c_\mu u, c_\mu Bu) = \left(\frac{B + B^*}{2}c_\mu u, c_\mu u\right) + \quad (8.34)$$

$$O(||u||_{t-1/2}^2) \quad \left(Re(v, Bv) = \frac{(v, Bv) + (Bv, v)}{2}\right).$$

The principal symbol of the commutator $[c_\mu, p_1]$ is equal to $\frac{1}{i}\{c_\mu, p_1\}$ and $[c_\mu, p_1] \in Op(\Sigma^t_{\sigma-1})$ uniformly with respect to μ. It can easily be seen that

$$Im(c_\mu u, [c_\mu, p_1]u) = Re(c_\mu u, \{c_\mu, p_1\}u) + O(||u||_{t-1/2}^2).$$

Applying Theorems 3.2, 3.3 from [24] we conclude that

$$Im(c_\mu u, [c_\mu, p_1]u) = Re\left(\frac{1}{2}\{c_\mu^2, p_1\}u, u\right) + O(||u||_{t-1/2}^2). \qquad (8.35)$$

Then combining Eq. (8.30), Eq. (8.31), Eq. (8.32), Eq. (8.33), Eq. (8.34) and Eq. (8.35) we get

$$\varepsilon||\eta c_\mu u||_0^2 + C(\varepsilon)||f||_{t,mcl(\Gamma)}^2 + d_1||u||_{t,mcl(\Gamma')}^2 \geq \qquad (8.36)$$

$$\frac{1}{2}Re(\{c_\mu^2, p_1\}u, u) + \frac{1}{2}Re\left(\left(B + B^* - \sum_{j=1}^n \frac{\partial A_j}{\partial x_j}\right)c_\mu u, c_\mu u\right) + O(||u||_{t-1/2}^2)$$

and the remainder does not depend on μ.

Using the identity Eq. (8.30) we have

$$\varepsilon||\eta c_\mu u||_0^2 + C(\varepsilon)||f||_{t,mcl(\Gamma)}^2 + d_2||u||_{t,mcl(\Gamma')}^2 \geq \qquad (8.37)$$

$$\frac{1}{2}Re(-\eta^2 c_\mu^2 Iu, u) + Re(Kc_\mu u, c_\mu u) + O(||u||_{t-1/2}^2),$$

as cone $supp((1 - \eta^2)A_{2t}) \subset\subset \Gamma'$, $\rho^0 \notin \Gamma'$.

The identity

$$Re(Kc_\mu u, c_\mu u) = Re(\eta^2 Kc_\mu u, c_\mu u) + Re((1 - \eta^2)Kc_\mu u, c_\mu u)$$

and standard results from [24] (see Corollary 3.5b) show that

$$Re(Kc_\mu u, c_\mu u) = Re(K\eta c_\mu u, \eta c_\mu u) + O(||u||_{t-1/2}^2 + ||u||_{t,mcl(\Gamma')}^2). \quad (8.38)$$

Obviously,

$$Re(\eta^2 c_\mu^2 Iu, u) = Re(I\eta c_\mu u, \eta c_\mu u) + O(||u||_{t-1/2}^2). \qquad (8.39)$$

Having in mind the relations Eq. (8.38), Eq. (8.39) we rewrite the a priori estimate Eq. (8.37) as:

$$\varepsilon||\eta c_\mu u||_0^2 + C(\varepsilon)||f||_{t,mcl(\Gamma)}^2 + d_3||u||_{t,mcl(\Gamma')}^2 \geq$$

$$Re\left(\left(-\frac{I}{2} + K\right)\eta c_\mu u, \eta c_\mu u\right) + O(||u||_{t-1/2}^2),$$

$\forall \varepsilon > 0$ and the reminder O does not depend on μ.

On the other hand $(-I/2 + K)$ is bounded in $\Sigma_{\sigma-1}^0$,

$$-\frac{I}{2} + K = \left(t - \frac{2\mu^2|\xi|^2}{1 + \mu^2|\xi|^2}\right) \cdot \left(\frac{K}{t - \frac{2\mu^2|\xi|^2}{1+\mu^2|\xi|^2}} + \sum_{j,k=1}^n \frac{\partial A_j}{\partial x_k}\frac{\xi_j \xi_k}{|\xi|^2}\right) \geq$$

$$(t-2) \cdot \left(\frac{K}{t} + \sum_{j,k=1}^{n} \frac{\partial A_j}{\partial x_k} \frac{\xi_j \xi_k}{|\xi|^2} \right) = \frac{t-2}{t} \left(K + t \cdot \sum_{j,k=1}^{n} \frac{\partial A_j}{\partial x_k} \frac{\xi_j \xi_k}{|\xi|^2} \right) > 0$$

in a conic ngbhd of ρ^0 according to $(iii)_t$ and $(ii)_0$.

In fact $\dfrac{K(x^0)}{t - \frac{2\mu^2 |\xi|^2}{1+\mu^2 |\xi|^2}} > \dfrac{K(x^0)}{t}$ as $K(x^0) > 0$. Taking cone $supp\eta$ in the same ngbhd of ρ^0 and applying Gärding's inequality for positive system of paradifferential operators (see [24]) we get that

$$Re \left(\left(-\frac{I}{2} + K \right) (\eta c_\mu u), \eta c_\mu u \right) \geq d_4 ||\eta c_\mu u||_0^2 + O(||\eta c_\mu u||_{-\gamma}^2),$$

for some $\gamma > 0$; $d_4 = const > 0$.

The classical interpolation inequality in Sobolev spaces

$$||w||_{-\gamma} \leq \delta ||w||_0 + C(\delta) ||w||_{-1}, \forall \delta > 0$$

enables us to conclude that for each $\mu \in (0, 1]$

$$||\eta c_\mu u||_0 \leq d_5,$$

i.e. $u \in H_{mcl}^t(\rho^0) \Rightarrow u \in H_{mcl}^t(\Gamma)$.

Thus Theorem 8.4 is proved.

8.4   Regularizing property of the solutions of a dissipative semilinear wave equation

In the paper [92] the authors considered the Cauchy problem to the following dissipative wave equation

$$\Box u + |u_t|^{h-1} u_t = 0, h > 1, \Box = \partial_t^2 - \Delta_x, \tag{8.40}$$

$$u|_{t=0} = f(x), u_t|_{t=0} = g(x), (t, x) \in \mathbf{R}^{1+n}, n \geq 2, u = u(t, x).$$

They proved the following global existence result.

Theorem 8.6.

Suppose that $\{f, g\} \in H^1(\mathbf{R}_x^n) \times L^2(\mathbf{R}_x^n)$, $H^s(\mathbf{R}_x^n)$ being the standard Sobolev space. Then there exists a unique solution u of Eq. (8.40) and such that $u \in C([0, \infty) : H^1(\mathbf{R}_x^n))$, $u_t \in C([0, \infty) : L^2(\mathbf{R}_x^n)) \cap L^{h+1}([0, \infty) \times \mathbf{R}_x^n)$. Moreover, two solutions of Eq. (8.40) whose Cauchy data agree on the ball $\{|x - x^0| \leq R\}$ must also agree on the truncated cone $\{|x - x^0| \leq R - t, 0 \leq t < T \leq R\}$.

The authors of [72] studied the problem Eq. (8.40) with f, g piecewise C^2, radial, compactly supported ($supp\{f, g\} \subset \{1 \leq |x| \leq R\}$) and having singularities on the sphere $\{|x| = 1\}$. Under some additional conditions imposed on $\{f, g\}$ they proved that the corresponding solution u is smoother in $\{t \geq 1\}$ than it is in the strip $\{0 \leq t < 1\}$. Similar results were shown in [90] where the assumption f, g to be radial was omitted.

We propose here a simple example illustrating the main theorem of [72]. It concerns the smoothing effect on the singularities developed for $t > 0$ and caused by the singularities existing for $t < 0$. In fact, let $n > 3$. Then the piecewise smooth solution $u \in C^0$ having finite jumps ∇u for $t < 0$ becomes C^1 smooth for $t > 0$. On the other hand, in the case $n = 3$ a solution u with logarithmic singularities for $t < 0$ develops C^1 regularity for $t > 0$.

Thus, in the nonlinear dissipative case we have regularizing effect on the singularities for $t > 0$, while in [106], $n = 2$, the interaction at 0 of the singularities carried out by 3 progressing waves (non-dissipative case) gave rise of new singularities for $t > 0$. They are weaker than the initial singularities in $\{t < 0\}$.

8.5 Formulation and investigation of the main dissipative nonlinear wave equation

1. Let $h = 2$, $n \geq 2$. Then by definition $u_\lambda(t, x) = u(\lambda t, \lambda x)$, $\forall \lambda > 0$ is a self-similar solution of Eq. (8.40). We seek now a radial in x self-similar solution of Eq. (8.40), i.e. a solution satisfying $u(t, r) = u(\lambda t, \lambda r)$, $r = |x|$, $x \in \mathbf{R}^n$ for each $\lambda > 0$. Put $\lambda = \frac{1}{r} > 0$. Then $u(t, r) = u(\frac{t}{r}, 1) = U(\frac{t}{r})$. As we know in spherical coordinates Eq. (8.40) takes the form

$$u_{tt} - u_{rr} - \frac{n-1}{r}u_r + u_t|u_t| = 0. \tag{8.41}$$

Put $s = \frac{t}{r}$. Then $U(s)$ satisfies the ODE

$$(1 - s^2)U'' + (n - 3)sU' + U'|U'| = 0. \tag{8.42}$$

Denote $V = U'$. The equation Eq. (8.42) is rewritten as:

$$(1 - s^2)V' + (n - 3)sV + V|V| = 0. \tag{8.43}$$

Having in mind that $|V|$ is Lipschitz continuous, we conclude that through each point $(s_0, V(s_0))$, $s_0 < -1$, $|s_0| < 1$, $s_0 > 1$ it is passing (at least locally) a unique smooth solution of Eq. (8.43).

Let $U = 0$ for $s < -1 \Rightarrow V = 0$ for $s < -1 \Rightarrow u = 0$ on the incoming cone $\{t \leq -r\} = \{t < 0, |t| \geq r\}$. We illustrate the case $n > 3$ with the following Fig. 8.5.

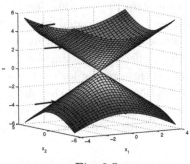

Fig. 8.5

Suppose now that $|s| < 1$ and make the change $V = (1 - s^2)^{\frac{n-3}{2}} W$ in Eq. (8.43). Obviously,

$$(1-s^2)W' + (1-s^2)^{\frac{n-3}{2}} W|W| = 0 \iff W' + (1-s^2)^{\frac{n-5}{2}} W|W| = 0. \quad (8.44)$$

The uniqueness theorem for Eq. (8.44) equipped with the condition $W(s_0) = 0$, $|s_0| < 1$ implies that $W(s) \equiv 0$. Therefore, $W(s)$ conserves its sign in its domain of existence contained in $\{|s| < 1\}$, i.e. if $W(s_0) < 0$ or > 0 then $W(s) < 0$ or > 0. Moreover, if W satisfies Eq. (8.44), then $-W$ is a solution of Eq. (8.46) too. Certainly, $W < (>)0$, implies that W is strictly monotonically decreasing (increasing).

2. We shall investigate below 3 different cases.

(A) $n > 3$, $|s| < 1$, $W > 0$. Then $\frac{1}{W(s)} - \frac{1}{W(s_0)} = G(s, s_0)$, where $W(s_0) > 0$ and $G(s, s_0) = \int_{s_0}^{s} (1 - \lambda^2)^{\frac{n-5}{2}} d\lambda$. The integral $G(s, s_0)$ can be calculated explicitly. For n-even we make the change $\lambda = \pm \frac{z}{\sqrt{1+z^2}}$, etc. Having in mind that $G'_s > 0$ for $|s| < 1$ and that $G(s, s_0)$ is convergent for $s = \pm 1$ we conclude that the strictly monotonically increasing function $G(s, s_0)$ satisfies the relations: $G(-1) \leq G(s) \leq G(1)$, $G(1) > 0$, $G(-1) < 0$, $G'(\pm 1) = 0$, for $n > 5$, $G'(\pm 1) = 1$ for $n = 5$ and $G'(\pm 1) = +\infty$ for $n = 4$. Therefore, $V(s) = \frac{(1-s^2)^{\frac{n-3}{2}} W(s_0)}{h(s,s_0)}$, where $h = 1 + W(s_0)G(s, s_0) = W(s_0)[\frac{1}{W(s_0)} + G(s, s_0)]$, $W(s_0) > 0$.

Depending on the location of $W(s_0) > 0$ on the positive half line we must consider 3 cases:

(A1) $\frac{1}{W(s_0)} + G(-1) > 0 \Rightarrow \frac{1}{W(s_0)} + G(s) > 0$, $|s| \leq 1 \Rightarrow V \in C^\infty(|s| < 1) \cap C(|s| \leq 1)$; $V(\pm 1) = 0$, $V(s) > 0$ for $|s| < 1$ and $n > 5 \Rightarrow V'(\pm 1) = 0$.

(A2) $\exists\ \bar{s}$, $|\bar{s}| < 1$ such that $G(\bar{s}, s_0) = -\frac{1}{W(s_0)} < 0$ and $G(s, s_0) > -\frac{1}{W(s_0)}$ for $s > \bar{s}$, $s_0 > \bar{s}$.

Certainly, then $0 > -\frac{1}{W(s_0)} > G(-1, s_0)$.

Evidently, V is defined in the interval $(\bar{s}, 1]$, $V(1) = 0$ and V blows up for $s = \bar{s}$

$(V(\bar{s} + 0) = +\infty)$.

(A3) $G(-1, s_0) + \frac{1}{W(s_0)} = 0$.

Then $V \in C^\infty(|s| < 1) \cap C(-1 < s \leq 1)$, $V(1) = 0$, $V(s) > 0$ for $|s| < 1$. Having in mind that $h(s, s_0) = W(s_0)R(s)$, where $R(s) = \int_{-1}^s (1 - \lambda^2)^{\frac{n-5}{2}} d\lambda$, $R(1) > 0$, we get that $V(s) = (1 - s^2)^{\frac{n-3}{2}}/R(s)$, $R(-1) = 0$. Applying L'Hospital rule we obtain that $V(-1) = n - 3 > 0$. So

$$U(s) = \begin{cases} 0, & s \leq -1 \\ \int_{-1}^s V(\lambda) d\lambda, & -1 < s \leq 1, U(s) \in C^0(\mathbf{R}) \cap C^1(\mathbf{R} \setminus \{-1\}) \\ E, & s > 1, E = \int_{-1}^1 V(\lambda) d\lambda > 0 \end{cases} \quad (8.45)$$

in case (A3). Thus, $U' = 0$ for $|s| > 1$; $U'(1 - 0) = V(1 - 0) = 0$, $V(s) \sim c(1 - s^2)^{\frac{n-3}{2}}$, $s \to 1 - 0$, $c = const > 0$, $U' = V$ has finite jump at $s = -1$.

Proposition 8.4. ([112])

The solution $u(t, r) = U(\frac{t}{r})$ of Eq. (8.40) in case (A3) is identically equal to $E > 0$ inside the characteristic light cone $\{t = r\}$, while $\nabla_{t,r} u \sim c|t - r|^{\frac{n-3}{2}}$ near $\{t = r\}$ (see Fig. 8.5). Moreover, $\nabla_{t,r} u$ has finite jump discontinuity at the characteristic cone $\{t = -r\}$. In case (A1) the function $U \in C^1(\mathbf{R}_s^1)$.

3. (B) $n = 3$, $|s| < 1$, respectively $s > 1$, $s < -1$.

In this case Eq. (8.43) takes the form

$$(1 - s^2)V' + V|V| = 0. \quad (8.46)$$

We know that V conserves its sign for $s < -1$, $s > 1$, $|s| < 1$. Again we must consider 3 cases:

(B1) $V(s_0) > 0$, $|s_0| < 1$. Then V is monotonically decreasing, $V(s) > 0$.

(B2) $s_0 < -1$, $V(s_0) > 0$. Then $V(s) > 0$ and $V(s)$ is monotonically increasing

(B3) $s_0 > 1$, $V(s_0) > 0$. Then $V(s) > 0$ and $V(s)$ is monotonically increasing.

We shall study in more details (B1) only, i.e. $|s| < 1$, $V(s) > 0$. After the separation of variables we have

$$\frac{1}{V(s)} - \frac{1}{V(s_0)} = \frac{1}{2}\left(log\frac{1+s}{1-s} - log\frac{1+s_0}{1-s_0} \right) \Rightarrow$$

$$V(s) = \frac{V(s_0)}{1 + \frac{1}{2}V(s_0)log(\frac{1+s}{1+s_0}\frac{1-s_0}{1-s})} \Rightarrow \tag{8.47}$$

$$V'(s) = \frac{1}{s^2 - 1}\frac{V^2(s_0)}{[1 + \frac{1}{2}V(s_0)log(\frac{1+s}{1+s_0}\frac{1-s_0}{1-s})]^2} < 0.$$

Letting $s \to 1 - 0$ we obtain $V(1 - 0) = +0$,

$$V(s) \sim \frac{-2}{log(1-s)}, V'(s) \sim \frac{2}{(s-1)log^2(1-s)}, s \to 1 - 0$$

According to L'Hospital rule $lim_{s\to1-0}V'(s) = -\infty$. So the monotonically decreasing function $V(s) > 0$, $1 > s > s_0$, $V(1) = 0$ possesses a vertical tangent at $s = 1 - 0$. One can easily see that $V(s)$ has a vertical asymptote for some $\bar{s} \in (-1,1)$, $\bar{s} < s_0$. In other words $1 + \frac{1}{2}V(s_0)log(\frac{1+s}{1+s_0}\frac{1-s_0}{1-s}) > 0$ for $s \in (\bar{s},1)$ and $1 + \frac{1}{2}V(s_0)log(\frac{1+s}{1+s_0}\frac{1-s_0}{1-s}) = 0$. To verify those two relations we put $k = e^{-\frac{2}{V(s_0)}} \Rightarrow 0 < k < 1$. Then $1 + \frac{1}{2}V(s_0)log(\frac{1+\bar{s}}{1+s_0}\frac{1-s_0}{1-\bar{s}}) = 0 \iff k\frac{1+s_0}{1-s_0} = \frac{1+\bar{s}}{1-\bar{s}}$, i.e. $\bar{s} = \frac{\beta-1}{\beta+1}$, where $\beta = k.\frac{1+s_0}{1-s_0} > 0$ as $|s_0| < 1$.

Evidently, $\beta > 0 \Rightarrow \bar{s} = \frac{\beta-1}{\beta+1} \in (-1,1)$ and that the mapping from $\{\beta > 0\}$ is onto $(-1,1)$.

In the case (B2) $lim_{s\to-\infty}V(s) = \frac{V(s_0)}{1+\frac{1}{2}V(s_0)log\frac{1-s_0}{1+s_0|}} = const > 0$. As in the previous case there exists a vertical asymptote of $V(s)$ for $\bar{s} > s_0$, $\bar{s} < -1$, $0 = h(\bar{s}) = 1 + \frac{1}{2}V(s_0)log|\frac{1+\bar{s}}{1+s_0}\frac{1-s_0}{1-\bar{s}}|$.

In the case (B3) $lim_{s\to1+0}V(s) = +0$, $g(s,s_0) = 1 + \frac{1}{2}V(s_0)log(\frac{s+1}{s_0+1}\frac{s_0-1}{s-1}) \Rightarrow \frac{\partial g}{\partial s} < 0$ for $s > 1$, $g(1) = \infty$, $g(s_0) = 1$, $lim_{s\to\infty}g = 1 + \frac{1}{2}V(s_0)log\frac{s_0-1}{s_0+1}$. Put $A(s_0) = lim_{s\to s_0}g$. Then if $A(s_0) \geq 0$ the solution $V(s)$ is defined for $s \geq 1$, while if $A(s_0) < 0$ then $V(s)$ exists on the interval $[1, \bar{s})$, where $g(\bar{s}, s_0) = 0$ and $lim_{s\to\bar{s}-0}V(s) = +\infty$; $1 < s_0 < \bar{s}$.

Geometrically, we have Fig. 8.6:

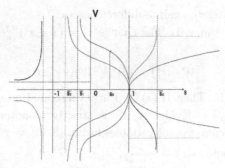

Fig. 8.6

Define now

$$U = \begin{cases} 0, & s < \bar{s} < 0 \\ \int_{s_0}^{s} V(\lambda)d\lambda, & \bar{s} < s < 1, \bar{s} < s_0 \\ E, & s \geq 1, \end{cases} \qquad (8.48)$$

where $E = \int_{s_0}^{1} V(\lambda)d\lambda > 0$.

Put $h(s) = 1 + \frac{1}{2}V(s_0)log(\frac{1+s}{1+s_0}\frac{1-s_0}{1-s})$, $|s| < 1$, $|s_0| < 1$, $-1 < \bar{s} < 0$, $\bar{s} < s$.

Then $h(s) = h(\bar{s}) + (s - \bar{s})h'(\bar{s}) + O(|s - \bar{s}|^2) = V(s_0)\frac{s-\bar{s}}{1-\bar{s}^2} + O(|s - \bar{s}|^2)$ as $h(\bar{s}) = 0 \Rightarrow U(s) \sim c_1 log(s - \bar{s})$, $s \to \bar{s} + 0$, $c_1 = const > 0$, i.e. $U(s) < 0$ for $\bar{s} < s < s_0$, $U(s_0) = 0$, $U(s) > 0$ for $1 \geq s > s_0$, $lim_{s \to 1-0}U(s) = E$, $U'(1 - 0) = V(1 - 0) = 0$.

Proposition 8.5. *(see* [112]*)*

The solution $u(t,r) = U(\frac{t}{r})$ of Eq. (8.40) in case (B) is given by Eq. (8.48), $U \in C^1(\mathbf{R}^1 \setminus \{\bar{s}\})$, U possesses logarithmic singularity near $\bar{s}(s > \bar{s})$, $U'(s) \sim \frac{-2}{log(1-s)}$ for $s \to 1 - 0$ and U is not continuous at $s = \bar{s} \in (-1, 0)$.

Therefore, the logarithmic singularity does not affect on U for $s > \bar{s}$ as U is C^1 smooth there.

Geometrically we have that $\nabla_{t,r}u$ is smooth outside and inside the non-characteristic light cone of the past $\{t = r\bar{s}\}$ but $|\nabla_{t,r}u(P)| = \infty$ for P tending to $\{t = r\bar{s}\}$ from outside. Moreover, $\nabla_{t,r}u \sim \frac{-2}{log|t-r|}$ for $t < r$, $t \approx r$.

4. (C), $n = 2$.

Then the equation Eq. (8.43) takes the form

$$(1 - s^2)V' - sV + V|V| = 0. \qquad (8.49)$$

We shall investigate again 3 different cases.

(C1): $|s| < 1$, $|s_0| < 1$. The change $V = (1 - s^2)^{-1/2}W$ in Eq. (8.49) leads to

$$W' + (1 - s^2)^{-3/2}W|W| = 0. \tag{8.50}$$

Let $W(s_0) > 0$. Then $W(s) > 0$ in its maximal interval of definition, contained in $(-1, 1)$. Evidently, W is strictly monotonically decreasing there. After the separation of variables in Eq. (8.50) we get $\frac{1}{W(s)} - \frac{1}{W(s_0)} = G(s) - G(s_0)$, where $G(s) = \frac{s}{\sqrt{1-s^2}}$, $|s| < 1$, $G'(s) = \frac{1}{(1-s^2)^{3/2}}$. Obviously, $G : (-1, 1) \to (-\infty, \infty)$ is a diffeomorphism, $G(-1 + 0) = -\infty$, $G(1 - 0) = \infty$, $G(0) = 0$, $G'(0) = 1$ and the inverse mapping G^{-1} is given by $s = \frac{G}{\sqrt{G^2+1}}$. Thus,

$$W(s) = \frac{W(s_0)}{1 + W(s_0)[G(s) - G(s_0)]} \tag{8.51}$$

is well defined for $1 > s \geq s_0$ and $W(1 - 0) = +0$.

Consider now the equation $G(\bar{s}) = G(s_0) - \frac{1}{W(s_0)} < G(s_0)$. Having in mind that G is strictly monotonically increasing we conclude that there exists a unique point $\bar{s} < s_0$ such that the function $j(s) = 1 + W(s_0)[G(s) - G(s_0)]$ has the properties: $j(s) > 0$ for $1 \geq s > \bar{s}$, $j(\bar{s}) = 0$ and consequently $lim_{s \to \bar{s}+0}W(s) = +\infty$. One can easily see that $lim_{s \to 1-0}W'(s) = -\infty$. In fact,

$$W' = -(1 - s^2)^{-3/2}W^2 = -(1 - s)^{-3/2} \cdot \frac{1}{[\frac{1}{W(s_0)} + (G(s) - G(s_0))]^2} \sim$$

$$-\frac{(1 - s^2)^{-3/2}}{G^2(s)} = -\frac{(1 - s^2)^{-1/2}}{s^2}$$

for $s \to 1 - 0$.

We are interested now in the behavior of $V(s) \in C^\infty((\bar{s}, 1)) \cup C^0((\bar{s}, 1])$ near the points $\bar{s}$ and 1.

Certainly, $j(s) = j(\bar{s}) + (s - \bar{s})j'(\bar{s}) + O(|s - \bar{s}|^2)$, $s \to \bar{s}$, i.e. $j(s) = k(s - \bar{s}) + O(|s - \bar{s}|^2)$, $s \to \bar{s} + 0$, $k = \frac{W(s_0)}{(1-\bar{s}^2)^{3/2}} > 0$.

Therefore,

$$V(s) = \frac{W(s)}{(1 - s^2)^{1/2}} \sim \frac{W(s_0)}{(1 - \bar{s}^2)^{1/2}k(s - \bar{s})} \tag{8.52}$$

for $s \to \bar{s} \in (-1, 1)$, i.e.

$$V(s) \sim \frac{1}{k_1(s - \bar{s})}, s \to \bar{s} + 0, k_1 = const > 0.$$

As we know,

$$V(s) = \frac{W(s_0)}{\sqrt{1-s^2}[1+W(s_0)(G(s)-G(s_0))]} = \frac{W(s_0)}{sW(s_0)+\sqrt{1-s^2}D},$$

(8.53)

where $D = 1 - \frac{s_0 W(s_0)}{\sqrt{1-s_0^2}} = 1 - G(s_0)W(s_0)$.

Thus, $lim_{s \to 1-0}V(s) = 1$.

Moreover, Eq. (8.53) implies

$$V'(s) = \frac{-W^2(s_0) + DW(s_0)\frac{s}{\sqrt{1-s^2}}}{(sW(s_0)+D\sqrt{1-s^2})^2}, |s_0| < 1, 1 \ge s > \bar{s}.$$

(8.54)

We must consider 3 different cases, namely

$$D > 0 \iff G(s_0)W(s_0) < 1. \text{ Then } lim_{s\to1-0}V'(s) = +\infty \quad (54-1)$$

$$D < 0 \iff G(s_0)W(s_0) > 1. \text{ Then } lim_{s\to1-0}V'(s) = -\infty \quad (54-2)$$

$$D = 0 \iff G(s_0)W(s_0) = 1. \text{ Then } lim_{s\to1-0}V'(s) = -1 \quad (54-3)$$

Evidently, $V'(s) \sim -1 + \frac{D}{\sqrt{2}W(s_0)}\frac{1}{\sqrt{1-s}}$ for $s \sim 1$.

Geometrical interpretation is given on Fig. 8.7.

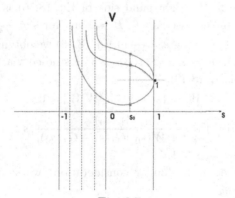

Fig. 8.7

Define now the function

$$U(s) = \begin{cases} 0, & s < \bar{s}, \quad -1 < \bar{s} < 0 \\ \int_{s_0}^s V(\lambda)d\lambda, & \bar{s} < s_0 < 1, \bar{s} < s \le 1 \\ E, & s > 1 \end{cases}$$

(8.55)

where $E = \int_{s_0}^{1} V(\lambda)d\lambda$ and V is defined as 0 for $s < \bar{s}$ and $s > 1$.

Evidently, $U \in C^1(\mathbf{R}^1 \setminus \{\bar{s}, 1\})$, U possesses logarithmic singularity for $s \to \bar{s} + 0$ (infinite jump), U is continuous for $s = 1$ but U' has a finite jump there: $U'(1 + 0) - U'(1 - 0) = -V(1 - 0) = -1$. In other words, the logarithmic singularity of U at $s = \bar{s}$ gives rise to a finite jump of U' at $s = 1$. The difference between the cases (B) and (C) is evident: U is smooth near $s = 1$ in the case (B), while U is continuous with finite jump of U' at $s = 1$ in the case (C).

Proposition 8.6. *(see [112])*

The solution $u(t, r)$ of Eq. (8.40) in case (C1) is continuous near the characteristic cone of the future $\{t = r\}$ but $\nabla_{t,r}u$ has a finite jump there, while $|\nabla_{t,r}^2 u| \sim |-1 + \frac{D}{\sqrt{2W(s_0)}} \frac{1}{\sqrt{|t-r|}}|$ for $t \to r$, $t < r$. Moreover, $u \sim \frac{1}{k_1}\log(t - r\bar{s})$ outside the non-characteristic cone $\{t = r\bar{s}\}$, $\frac{t}{r} \to \bar{s} + 0$.

Remark 8.5. We can propose an explicit formula for Eq. (8.55). In fact, $\int V(\lambda)d\lambda = \int \frac{W(s_0)ds}{sW(s_0)+D\sqrt{1-s^2}}$, $W(s_0) > 0$. One can see that

$$\int \frac{ds}{As + B\sqrt{1-s^2}} = \frac{1}{A^2 + B^2}(Barcsin\, s + Alog|As + B\sqrt{1-s^2}|), \quad (8.56)$$

where $A > 0$, $B > 0$ or $A > 0$, $B < 0$, $|s| < 1$.

The primitive in the right-hand side of Eq. (8.56) is well defined for $s \neq \frac{-B}{\sqrt{A^2+B^2}} < 0$ in the case $A > 0$, $B > 0$ and for $s \neq \frac{B}{\sqrt{A^2+B^2}} > 0$ in the case $A > 0$, $B < 0$. Putting $s = \frac{t}{|x|}$ in Eq. (8.56) we obtain our solution U.

The case (C2): $n = 2$, $s > 1$, $s_0 > 1$ is studied via the change $V = (s^2 - 1)^{-1/2}W$ in Eq. (8.49). Then

$$W' - (s^2 - 1)^{-3/2}W|W| = 0, \quad (8.57)$$

$$W(s) = \frac{W(s_0)}{1 + W(s_0)[G_1(s) - G_1(s_0)]}, s > 1$$

and $G_1(s) = -\frac{s}{\sqrt{s^2-1}}$, $s > 1$.

We omit the details. Similar considerations work in the case (C3): $n = 2$, $-s > 1$, $-s_0 > 1$.

Examples: 1) Consider Eq. (8.40), case (A3) with $n = 5$.

Then $U(s) = \begin{cases} 0, & s \leq -1 \\ 2 - \frac{(s-1)^2}{2}, & -1 < s \leq 1 \\ 2, & s > 1 \end{cases}$. Putting $s = \frac{t}{r}$

we get $u(t, x) = \begin{cases} 0, & t \leq -r \\ \frac{3r^2+2rt-t^2}{2r^2}, & |t| \leq r, r = |x| \\ 2, & t \geq r \end{cases}$ (see Fig. 8.8).

2) $u(t,x) = \frac{1}{2}(arcsin\frac{t}{r} + log(\frac{t}{r} + \sqrt{1 - \frac{t^2}{r^2}}))$, $\frac{t}{r} \neq -\frac{1}{\sqrt{2}}$, say $1 > \frac{t}{r} > -\frac{1}{\sqrt{2}}$ (case $(C1)$) (see Fig. 8.9).

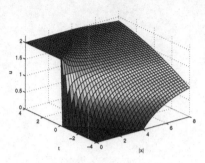

Fig. 8.8

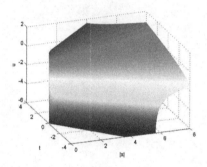

Fig. 8.9

The computations in Section 8.5 are elementary and the readers will understand them better.

Bibliography

[1] Abbott, M.R. (1956). A theory of the propagation of bores in channels and rivers, *Proc. Camb. Phil.Soc.* **52**, pp. 344–362.

[2] Ablowitz, M. and Baldwin, D. (2012). Nonlinear shallow ocean wave soliton interactions on flat beaches, *ArXiv:1208.2904 v1 [nlin.PS]*.

[3] Ablowitz, M. and Clarkson, P. (1991). *Solitons. Nonlinear evolution equations and Inverse Scattering*, (Cambridge University Press).

[4] Ablowitz, M., Ramani, A. and Segur, H. (1980). A connection between nonlinear evolution equations and ordinary differential equations of p-type, I, II, *J. Math. Phys.* **21**, a), b), pp. 715–721, pp. 1006–1015.

[5] Ablowitz, M. and Segur, H. (1981). *Solitons and the inverse scattering transform*, (SIAM Philadelphia).

[6] Abramowitz, M. and Stegun, I. (1972). *Handbook of mathematical functions*, (Dover, New York).

[7] Aero, E.L., Buligin, A. and Pavlov, Yu. (2009). Solutions of the 3 dimensional sin-Gordon equation, *Theoret. and Math.Physics* **158:3**, pp. 370–377.

[8] Akhiezer, N. (1990). *Elements of the theory of elliptic functions*, (Translations of Mathematical Monographs) **79** (AMS).

[9] Akhmediev, N., Ankiewics, A. and Taki, M. (2009). Waves that appear from nowhere and disappear without a trace, *Physics Letters A* **373**, pp. 675–678.

[10] Akhmediev, N., Eleonskii, V. and Kulagin, N. (1985). Generation of periodic trains of picosecond pulses in an optical fiber: exact solutions, *Zh. Eksp. Teor. Fiz.* **89**, pp. 1542–1551.

[11] Akhmediev, N. and Korneev, V. (1986). Modulation instability and periodic solutions of the nonlinear Schrödinger equation, *Theor. Math. Ph.* **69:2**, pp. 189–194.

[12] Alinhac, S. and Gérard, P. (1991). *Opérateurs pseudo-differentiels et théoréme de Nash-Moser*, (Lectures, InterEditions /Editions du CNRS, Paris).

[13] Arnold, V. (1992). *Ordinary differential equations*, (Springer).

[14] Arnold, V.I. (1978). *Additional chapters on the theory of ordinary differential equatios*, (Nauka, Moscow).

[15] Bäcklund, A. (1880). Zür Theorie dell partiellen Differential gleichungen erster Ordnung, *Math. Ann.* **27**, pp. 285–328.

[16] Beals, M. (1982). Spreading of singularities for a semilinear wave equation, *Duke Math. J.* **49**, pp. 275–286.

[17] Beals, M. (1984). Nonlinear wave equation with data singular at one point, *Cont. Math.* **27**, pp. 83–95.

[18] Beals, M. (1987). Interaction of radially smooth nonlinear waves, *Lecture notes in Math.* **1256**, pp. 1–27.

[19] Beals, R., Sattinger, D. and Szmigielski, J. (2001). Peakon-antipeakon interaction, *J. of Nonlinear Math. Ph.* **8**, pp. 23–27.

[20] Ben-gong Zhang, Liu, Z. and Xiao, Q. (2010). New exact solitary wave and multiple soliton solutions of quantum Zakharov-Kuznetsov equation, *Appl. Math. and Computation* **217**, pp. 392–402.

[21] Bhatnagar, P.L. (1979). *Nonlinear waves in one dimensional dispersive systems*, (Oxford Math. monograph).

[22] Bishop, R. (1975). There is more than one way to frame a curve, *Amer. Math. Monthly* **82**, pp. 246–251.

[23] Bony, J.M. (1979). Interaction des singularités pour les équations aux deriveés partielles nonlineaires, *Sem. Goulaouic-Meyer-Schwartz* exp. No. 22.

[24] Bony, J.M. (1981). Calcul symbolique et propagation des singularités pour les équations aux dérivées partielles nonlineaires, *Ann. Scien. de l'Ecole Norm. Sup.* **14**, pp. 209–246.

[25] Bony, J.M. (1984). Second microlocalization and propagation of singularities for semilinear hyperbolic equations, *Tanikachi Symp. Katatá*, pp. 11–49.

[26] Boussinesq, J. (1871). Théorie de l'intumescence liquide appelée onde solitaire où de translation se propageant dans un canal rectangulaire, *Comptes Rendus* **72**, pp. 755–759.

[27] Boyd, R. (1992). *Nonlinear optics*, (Academic Press, NY).

[28] Budanov, V. (2013). On an isochronal nonlinear system, *Vestnik of Moscow Univ., Series Math. and Mech* No. 6, pp. 59–63.

[29] Bullough, R. and Caudrey, P. eds. (1980). *Solitons*, (Springer-Verlag, Berlin-New York).

[30] Byrd, P. and Friedman, M. (1971). *Handbook of elliptic integrals for engineers and scientists*, (NY, Springer-Verlag).

[31] Camassa, R. and Holm, D. (1993). An integrable shallow water equation with peaked solitons, *Phys. Rev. Lett.* **71**, pp. 1661–1664.

[32] Caudrey, P., Gibbon, J., Eilbock, J. and Bullough, R. (1973). Exact multisoliton solutions of self-induced transparency and sine-Gordon equations, *Phys. Rev. Lett.* **30**, pp. 237–238.

[33] Chandrasekharan, K. (1985). *Elliptic functions*, (Springer, Berlin).

[34] Chen, S. (2011). *Analysis of singularities for partial differential equations*, (World Scientific).

[35] Coddington, E. and Levinson, N. (1955). *Theory of ordinary differential equations*, (McGraw-Hill, NY).

[36] Constantin, A. and Esher, J. (1998). Global existence and blow up for a

shallow water equation, *Ann. Sc. Norm. Sup. Pisa, IV Ser.* **26**, pp. 303–328.

[37] Cotter, C., Holm, D., Ivanov, R. and Parcival, J. (2011). Waltzing peakons and compacton pairs in a cross-coupled Camassa-Holm equation, *Journal of Phys. A: Math. Theor.* **44**, 265205.

[38] Courant, R. (1962). *Partial differential equations* **II** (Interscience publishers, NY).

[39] Curtright, T. and Fairlie, D. (2002). Extra dimensions and nonlinear equations, *arXiv:math-ph/0207008v1*, pp. 1–8.

[40] Darboux, G. (1915). *Lecons sur la Theórie générale des surfaces et les applications géometriques du calcul infinitésimal* 2 (Gauthier Villars, Paris).

[41] Degasperis, A. and Procesi, M. (1999). Asymptotic integrability, symmetry and perturbation theory *in: A. Degasperis and G. Gaeta (Eds.)*, (World Scientific, Singapore), pp. 23–37.

[42] Drazin, P. and Johnson, R. (1989). *Solitons: An Introduction*, (Cambridge Univ. Press).

[43] Dressler, R.F. (1949). Mathematical solutions of the problem of roll waves in inlined open channels, *Comm. Pure Appl. Math.* **2**, pp. 140–190.

[44] Dwight, H. (1961). *Tables of integrals and other mathematical data*, (Mcmillan company, NY).

[45] Fedorjuk, M. (1983). *Asymptotical methods for linear ordinary differential equations*, ("Nauka", M.) (in Russian).

[46] Fokas A. (1995). On a class of physically important integrable equations, *Physica D* **87**, pp. 145–150.

[47] Fornberg, B. and Whitham, G. (1978). A numerical and theoretical study of certain nonlinear wave phenomena, *Philos. Trans. R. Soc. London* **ser. A 289**, pp. 373–404.

[48] Fuchssteiner, B. (1996). Some tricks from the symmetry-toolbox for nonlinear equations: Generalizations of the Camassa-Holm, *Phys. D* **95**, pp. 229–243.

[49] Gaeta, G., Gramchev, T. and Walcher, S. (2007). Compact solitary waves in linearly elastic chains with non–smooth on–site potential, *J. Phys. A: Math. Theor.* **40**, pp. 4493–4509.

[50] Georgiev, V., Tzvetkov, N. and Visciglia, N. (2013). On the continuity of the solution map for the cubic 1d periodic NLW equation, preprint, *Univ. of Pisa*.

[51] Gérard, P. and Grellier, S. (2010). The cubic Szegö equation, *Ann. Sci. de L'ENS* **43:5**, pp. 761–810.

[52] Gerdjikov, V., Vilasi, G. and Yanovski, A. (2008). *Inegrable Hamiltonian Hierarchies. Spectral and Geometrical Methods*, (Lecture Notes in Physics) **748**, Springer.

[53] Gerdjikov, V. and Stefanov, A. (2017). New examples of two-component NLS equations, *Report on the conference NTADES 2017* Sofia.

[54] Goritskii, A., Krujkov, S. and Chechkin, G. (1997). *Quasilinear first order partial differential equations*, (Moscow State Univ. Lomonosov ed., Moscow).

[55] Gradsteyn, M. and Ryzhik, I. (1980). *Tables of integrals, series and products*, (Academic Press, NY-London).

[56] Grakhov, F. (1966). *Boundary value problems*, (Pergamon Press, Oxford).

[57] Guo, Y. and Lai, Sh. (2010). New exact solutions for an $(N+1)$-dimensional generalized Boussinesq equation, *Nonlinear analysis* **72**, pp. 2863–2873.

[58] Hasimoto, H. (1972). A soliton on a vortex filament, *J. Fluid Mech.* **51**, pp. 477–485.

[59] Hereman, W. (1993). Symbolic software for the study of nonlinear partial differential equations, *Advances in Computer methods for PDE VII, IMACS*, pp. 326–332.

[60] Hereman, W. and Nuseir, A. (1997). Symbolic methods to construct exact solutions of nonlinear partial differential equations, *Math. Comp. Simul.* **43**, pp. 13–27.

[61] Hirota, R. (1972). Exact solution of the modified Korteweg de Vries equation for multiple collisions of solitons, *J. Phys. Soc. Japan* **33**, pp. 1456–1458.

[62] Hirota, R. (1972). Exact solution of the sine-Gordon equation for multiple collisions of solitons, *J. Phys. Soc. Japan* **33**, pp. 1459–1463.

[63] Hirota, R. (1976). Direct method of finding exact solutions of nonlinear evolution equations, *Lecture Notes in Math.* **515**, (Berlin, Springer), pp. 41–68.

[64] Hirota, R. (1980). Direct methods in soliton theory, *in Solitons, eds. R. Bullough and P. Caudrey, Topics in Current Physics, Springer* **17**, pp. 174–192.

[65] Hirota, R. (2004). *The direct method in soliton theory*, (Cambridge University Press).

[66] Hörmander, L. (1963). *Linear partial differential operators*, (Springer-Verlag, Berlin).

[67] Hörmander, L. (1985). *The analysis of linear partial differential operators III, IV*, (Springer-Verlag, Berlin-Heidelberg).

[68] Inc, M. (2008). New solitary wave solutions with compact support and Jacobi elliptic function solutions for the nonlinearly dispersive Boussinesq equation, *Chaos, Solution, Fractals* **37**, pp. 792–798.

[69] Infeld, E. and Rowlands, G. (2000). *Nonlinear waves, solitons and chaos*, (Cambridge Univ. Press, Cambridge).

[70] Ivanov, R. (2005). On the integrability of a class of nonlinear dispersive equations, *J. Nonlinear Math. Ph.* **1295**, pp. 462–468.

[71] Ivanov, R., Lyons, T. and Orr, N. (2017). A dressing method for soliton solutions of the Camassa-Holm equation, *arXiv:1702.01128 V1 [nlin. SI]*.

[72] Joly, J.-L., Métivier, G. and Rauch, J. (2000). Nonlinear hyperbolic smoothing at a focal point, *Michigan Math. J.* **47:2**, pp. 295–312.

[73] Kadomtsev, B. and Petviashvili, V. (1970). On the stability of solitons in a medium with weak dispersion, *Soviet Math. Doklady* **192**, pp. 753–756.

[74] Kamke, E. (1959). *Differentialgleichungen*, **I**, (Gewöhliche Differentialgleichungen, Leipzig).

[75] Kaptsov, O. and Shanko, Ju. (1999). Multiparametric solutions of Tzitzeika equation, *Diff. Eq.* **35:12**, pp. 1660–1668.

[76] Kärtner, F., Kopf, D. and Keller, U. (1994). Solitary pulse stabilization and shortening in actively mode-locked lasers, *J. Opt. Soc. Amer. B Opt. Phys.* **12**, pp. 486–496.

[77] Kaup, D. (1980). On the inverse scattering problem for the cubic eigenvalue problems for the class $\psi_{3x} + 6Q\psi_x + 6R\psi = \lambda\psi$, *Stud. Appl. Math.* **62**, pp. 189–216.

[78] Knobel, R. (2000). An introduction to the mathematical theory of waves, *Student Math. Library*, **vol. 3**, AMS.

[79] Kohlenberg, J., Lundmark, H. and Szmigielski, J. (2007). The inverse spectral problem for the discrete cubic string, *Inverse problems* **23**.

[80] Kolkovska, N. and Dimova, M. (2012). A new conservative finite difference scheme for Boussinesq paradigm equation, *Central European, J. of Math.* **10**, pp. 1159–1171.

[81] Kostov, N., Gerdjikov, V. and Valchev, T. (2010). Multicomponent nonlinear systems of Bosé-Fermi fields: Exact solutions, *AIP Conference Proc.* **1340**, *Int. Workshop on Complex structures, Integrability and Vector fields, Sofia*, pp. 111–125.

[82] Kudryashov, N. and Sinelnikov, D. (2012). Exact solutions of the Swift-Hohenberg equation with dispersion, *Comm. Nonlinear Sci. Numer. Simul.* **17**, pp. 26–30.

[83] Kudryashov, N. and Ryabov, P. (2016). Analytical and numerical solutions of the generalized dispersive Swift-Hohenberg equation, *Appl. Math. and Computation* **286**, pp. 171–177.

[84] Kuperschmidt, B. (1984). A super KdV equation: an integrable system, *Phys. Lett.* **102 A**, pp. 213–215.

[85] Kutev, N., Kolkovska, N. and Dimova, M. (2014). Global existence to generalized Boussinesq equation with combined power type nonlinearities, *J. Math. An. and Appl.* **410**, pp. 427–444.

[86] Kutz, J.N. (2006). Mode-locked soliton lasers, *SIAM review* **48:4**, pp. 629–678.

[87] Kutz, J., Collings, B., Bergman, K. and Knox, W. (1998). Stabilized pulse spacing in soliton lasers due to gain depletion and recovery, *IEEE J. Quant. Elec.* **34**, pp. 1749–1757.

[88] Lamb, G. (1980). *Elements of soliton theory*, (John Wiley and Sons, NY).

[89] Lax, P. (1968). Integrals of nonlinear equations of evolution and solitary waves, *Comm. Pure Appl. Math.* **21**, pp. 467–490.

[90] Liang, J. (2009). Hyperbolic smoothing effect for semilinear wave equations at a focal point, *Journal of Hyperb. Diff. Equations* **6:1**, pp. 1–23.

[91] Li, Ji-bin (2008). Exact traveling wave solutions to 2D-generalized Benney-Luke equation, *Appl. Math. Mech.* **29:11**, pp. 1391–1398.

[92] Lions, J.L. and Strauss, W. (1965). Some nonlinear evolution equations, *Bull. Soc. Math. Fr.* **93**, pp. 43–96.

[93] Lundmark, H. (2007). Peakons and shock peakons in the Degasperis-Procesi equation, *NEEDS 2007, L'Ametlla de Mar*, pp. 1–33.

[94] Lundmark, H. and Szmigielski, J. (2003). Multipeakon solutions of the Degasperis-Procesi equation, *Inverse problems* **19**, pp. 1241–1250.

[95] Manakov, S. (1974). On the theory of two-dimensional stationary self-focusing of elektromagnetic waves, *Sov. Phys. JETP* **38**, pp. 248–253.

[96] Matsuno, Y. (2013). Smooth multisoliton solutions and their peakon limit of Novikov's Camassa-Holm type equation with cubic nonlinearity, *ArXiv:1305-6728 v1 [nlin.SI]*.

[97] Morrison, A., Parkes, E. and Vakhnenko, V. (1999). The N loop soliton solution of the Vakhnenko equation, *Nonlinearity* **12**, pp. 1427–1437.

[98] Moser, J. (1966). A rapidly convergent iteration method and nonlinear differential equations I, II *Ann. Sc. Norm. Sup. Pisa III* **2:20**, pp. 265–315, **3:20**, pp. 499–535.

[99] Moslem, W., Ali, S., Shukla, P., Tang, X. and Rowlands, G. (2007). Solitary, explosive and periodic solutions of the quantum Zakharov-Kuznetsov equation and its transverse instability, *Phys. Plasmas* **14**, 082308-1-082308-5.

[100] Müller, P., Garrett, Ch. and Osborne, A. (2005). Rogue waves, in: The 14th Aha Huliko' a Hawaiian Winter Workshop, *Oceanography* **18:3**, 66.

[101] Muskhelischvili, N. (1958). *Boundary value problems of functions theory and their applications to Mathematical Physics*, (Wolters-Noorhoff Publisher, Gröningen).

[102] Newell, A. (1985). *Solitons in Mathematics and Physics*, (SIAM).

[103] Olver, P. and Rosenau, P. (1996). *Phys. rev. E* **53**, pp. 1900–1906.

[104] O'Neil, J., Kutz, J. and Sandstede, B. (2002). Analytic theory for actively mode-locked fiber lasers, *IEEE J. Quant. Elec.* **38**, pp. 1412–1419.

[105] Popivanov, P. (1993). Wave fronts of the solutions of fully nonlinear symmetric positive systems of partial differential equations, *Boll. Unione Mat. It* **7-B**, pp. 643–652.

[106] Popivanov, P. (2003). Nonlinear PDE. Singularities, propagation, applications in Nonlinear Hyperbolic equations, *Spectral theory and Wavelet transformations, Birkhäuser series, Operator theory-Advances and Applications* **148**, pp. 1–94.

[107] Popivanov, P. (2006). *Geometrical methods for solving of fully nonlinear partial differential equations*, (Mathematics and Its Applications Series, Ed. of the Union of Bulg. Mathematicians, Sofia).

[108] Popivanov, P. (2011). Traveling waves for some generalized Boussinesq type equations, *Int. workshop on complex structures, integrability and vector fields, AIP*, **1340**, pp. 126–134.

[109] Popivanov, P. (2014). Explicit formulas to the solutions of several equations of mathematical physics, *Pliska* **23**, pp. 24–36.

[110] Popivanov, P. and Slavova, A. (2011). *Nonlinear waves. An introduction*, (World Scientific, NY, London, Singapore; Series on Analysis, Appl. and Computation) **4**.

[111] Popivanov, P. and Slavova, A. (2012). Full classification of traveling wave solutions of Fornberg-Whitham equation. Solutions into explicit form, *C. R. Acad. Bulg. Sci.* **65:5**, pp. 563–574.

[112] Popivanov, P., Slavova, A. and Zecca, P. (2010). Regularizing property of the solutions of dissipative semilinear wave equation, *C. R. Acad. Bulg. Sci.* **63:7**, pp. 961–970.

[113] Proctor, J. and Kutz, J. (2006). Theory of Q-switching in actively mode-locked lasers, *J. Opt. Soc. Amer. B Opt. Physics* **23**, pp. 652–662.

[114] Prössdorf, S. (1974). *Einige klassen singulärer gleichingen*, (Math. Mono-grphien), **Band 37**, (Akademie Verlag, Berlin).

[115] Qiao, Z. (2006). A new integrable equation with cuspons and W/M-shape-peaks solitons, *J. Math. Phys.* **48**, 112701.

[116] Qiao, Z., Xia, B. and Li, J. (2012). Integrable system with peakon, weak kink, and kink-peakon interactional solutions, *arXiv:1205.2028 v2 [nlin.SI]*.

[117] Rauch, J. (2012). *Hyperbolic partial differential equations and geometric optics*, (Graduate studies in Math.) **133** AMS.

[118] Rauch, J. and Reed, M. (1980). Propagation of singularities for semilinear hyperbolic equations in one space variable, *Ann. of Math.* **111**, pp. 531–552.

[119] Rauch, J. and Reed, M. (1985). Striated solutions of semilinear, two speed wave equations, *Indiana Univ. Math. J.* **34**, pp. 337–353.

[120] Rijik, I. and Gradstein, M. (1980). *Table of integrals, series and products*, (Academic Press).

[121] Rozhdestvenskij, B. and Yanenko, N. (1983). Systems of quasilinear equations and their applications to gas dynamics, *Transl. Math. Monographs* **55**, (AMS, Providence, RI).

[122] Satsuma, J. (1976). N-soliton solution of the 2-dimensional Korteweg de Vries equation, *J. Phys. Soc. Japan* **40**, pp. 286–290.

[123] Satsuma, J. and Kaup, D. (1977). A Bäcklund transformation for a higher order KdV equation, *J. Phys. Soc. Japan* **43**, pp. 692–697.

[124] Sawada, K. and Kotera, T. (1974). A method of finding N-soliton solutions of the KdV and KdV-like equation, *Prog. Theor. Ph.* **51**, pp. 1355–1367.

[125] Shaw, J. (2004). *Mathematical principles of optical fiber communications*, (SIAM, Philadelphia).

[126] Swift, J. and Hohenberg, P. (1977). Hydrodynamic fluctuations at the convective instability, *Phys. Rev. A* **15**, pp. 319–328.

[127] Taylor, M. (1981). *Pseudodifferential operators*, (Princeton Univ. Press, Princeton, N.J.).

[128] Tzitzeika, G. (1910). Sur une nouvelle classe de surfaces, *C. R. Acad. Sci. Paris* **150**, pp. 955–956.

[129] Tzitzeika, G. (1910). *Ibid* **150**, pp. 1227–1229.

[130] Vakhnenko, V. (1992). Solitons in a nonlinear model medium, *J. Phys. A: Mat. gen.* **25**, pp. 4181–4187.

[131] Vakhnenko, V. and Parkes, E. (1998). The two loop soliton solution of the Vakhnenko equation, *Nonlinearity* **11**, pp. 1457–1464.

[132] Valchev, T. (2016). On soliton equations in classical differential geometry, *Pliska* **26**, pp. 253–262.

[133] Vekua, N. (1967). *Systems of singular integral equations*, (P. Noordhoff Ltd., Gröningen).

[134] Vitanov, N., Dimitrova, Z. and Kantz, H. (2013). Application of the method

of simplest equation for obtaining exact traveling wave solutions for the extended Korteweg-de Vries equation and generalized Camassa-Holm equation, *Appl. Math. and Computation* **219**, pp. 7480–7492.

[135] Vladimirov, V. (1971). *Equations of Mathematical Physics*, (Volume 3 of Pure and Applied Mathematics Series).

[136] Wazwaz, A. (2007). Multiple-soliton solutions for the Boussinesq equation, *Appl. Math. and Computation* **192**, pp. 479–486.

[137] Wazwaz, A. (2008). New traveling wave solutions to the Boussinesq and the Klein–Gordon quations, *Comm. Nonlinear Sci., Numer. simul.* **13**, pp. 889–901.

[138] Whitham, G. (1974). *Linear and nonlinear waves*, (A Willey-Interscience Publ.).

[139] Whittaker, E. and Watson, G. (1927). *A course of modern analysis*, (Cambridge Univ. Press, Cambridge).

[140] Yan, Z. (2007). Similarity transformations and exact solutions for a family of higher-dimensional Boussinesq equations, *Phys. Lett. A* **361**, pp. 223–230.

[141] Zabuski, N.J. (1967). In: *Nonlinear waves* (ed. W. Ames), *New York, Academic Press*, pp. 223–258.

[142] Zakharov, V. and Kuznetsov, E. (1974). On three dimensional solitons, *Sov. Phys. JETP* **39**, pp. 285–288.

[143] Zakharov, V. and Manakov, S. (1973). Resonant interaction of wave packets in nonlinear media, *Soviet Phys. JETP Lett.* **18**, pp. 243–245.

[144] Zakharov, V., Manakov, S., Novikov, S. and Pitaevskii, L. (1984). *Theory of solitons: The inverse scattering method*, (Plenum, NY).

[145] Zakharov, V. and Shabat, B. (1972). Exact theory of two-dimensional self focusing and one-dimensional self modulation of waves in nonlinear media, *Sov. Phys. JETP* **34**, pp. 62–69.

[146] Zakharov, V. and Shabat, A. (1973). Interaction between solitons in a stable medium, *Sov. Phys. JETP* **37**, pp. 823–827.

[147] Zakharov, V. and Shabat A. (1974). A scheme for integration of nonlinear evolution equations of Mathematical Physics by the method of inverse scattering transform I, *Funct. An. and Appl.* **6:3**, pp. 43–53.

[148] Zhou, J.B. and Tian, L.X. (2010). Solitons, peakons and periodic cusp wave solutions for Fornberg-Whitham equation, *Nonlinear Analysis: RWA* **11**, pp. 356–363.

[149] Zhu, Y. and Lu, C. (2007). New solitary solutions with compact support for Boussinesq-like $B(2n, 2n)$ equations with fully nonlinear dispertion, *Chaos, Solution, Fractals* **32**, pp. 768–772.

Index

Printed in the United States
By Bookmasters